CONTRIBUTING AUTHORS

Robert H. Coffin, Thilmany Pulp and Paper Company, Kaukauna, Wis.

Albert D. Gusman, President, Gusman Associates, New York, N. Y.

Robert A. A. Hentschel, Research Manager, Spunbonded Division, E. I. du Pont de Nemours and Co., Inc., Wilmington, Dela.

Jack M. Rudolph, Manager, Market Research, Pulp and Paper Division, Kimberly-Clark Corporation, Neenah, Wis.

Michael C. Sacher, Consumer Marketing Research Division, Kimberly-Clark Corporation, Neenah, Wis.

John C. Williams, Vice President for Research, Cuno Division, American Machine and Foundry Co., Meriden, Conn.

INDUSTRIAL AND SPECIALTY PAPERS

Volume IV—Product Development

Prepared by a Staff of Specialists
Under the Editorship of

Robert H. Mosher

R. M. Associates
Neenah, Wisconsin

and

Dale S. Davis

Professor Emeritus, Pulp and Paper Technology
University of Alabama

CHEMICAL PUBLISHING COMPANY, INC.
New York   1970

Industrial and Specialty Papers, Volume 4

ISBN: 978-0-8206-0223-3

Chemical Publishing Company:
www.chemical-publishing.com
www.chemicalpublishing.net

First edition:
© **Chemical Publishing Company, Inc.** – New York, 1970
Second Impression:
Chemical Publishing Company, Inc. - 2011

Printed in the United States of America

FOREWORD

The first three volumes of this work deal with various aspects of the technology, manufacture, and application of industrial and specialty papers. This is a broad and complex subject, and the editors have treated the matter in considerable depth. Emphasis has been properly placed on the applications of these papers, and particularly on the end use of the products produced.

By definition, specialty papers are manufactured to meet unique requirements. The special techniques of paper manufacture and converting described in previous volumes were generally developed to meet particular market specifications.

Volume IV considers the business, product development, and marketing aspects of this industry, and describes three important new products that are making significant changes in the paper industry. The very nature of the products described in the first three volumes requires at least some consideration of the areas of demand that created the incentive to develop these processes and techniques.

Volume IV also gives important consideration to proved techniques of product development and market research that have developed so rapidly in the paper industry during the past ten years.

One fact is evident, namely, that the paper industry can no longer rely on sales departments to sell only what it can make. Those companies moving ahead in our industry are the ones that have realized that maximum effectiveness in the market place comes from proper application of the concept that industry must produce what the market requires. Indeed, success is becoming more and more dependent on the ability of manufacturing facilities to adapt to rapid changes. Our industry is becoming increasingly conscious of the fact that it must pay particular attention to the function of marketing within our customers' rapidly changing technological environment.

The ratio of capital investment to unit sales in the paper industry is very high, and consequently, the risk attendant upon the utiliza-

i

tion of capital to develop new markets and satisfy new demands is considerable. The paper industry can no longer afford to handle product development on a "hit-or-miss" basis. The authors, in this volume, describe a number of tools available to management which, when properly applied, minimize risk and maximize the probability of success. The techniques described in Volume IV can provide management with a new tool, that of creating an extrapolation into the future technological requirements of the consuming areas. Market research and product development must be more than just a source of new products and new ideas. They actually can provide the discerning company with a road map showing how to move from one point to another with a considerable degree of accuracy.

It will not be enough for the paper company of the future to rely on its product development and market research departments to tell of demands that already exist. They will have to develop the faculty of anticipating market requirements and have the products ready as the demand for them is created in the market place. Certainly the application of the computer to the evaluation of product development techniques and the ability of the computer to make accurate business evaluations and analyses will change product development concepts in the paper industry.

Some concepts used by the specialty and industrial paper industry today will become mandatory for the commodity producers of paper tomorrow. In this respect, the significance of this series is much broader than might be indicated by the title.

This volume and its companions are unique in their treatment of the specialty and industrial paper fields. They also fill a long recognized need in the paper industry. The authors are to be complimented on the excellent job they have done in defining the parameters of this broad and unique area.

R. T. Seith
Vice President, Marketing
Mosinee Paper Mills Company

Contents

chapter 1

New Product Development —
An Overview

MICHAEL C. SACHER

I. INTRODUCTION

Few areas of American business today are as exciting as "new product development." Mention the three magic words to the company president or the most recently hired employee, and the eyes of each light up with interest.

By whatever name—interest—curiosity—glamour—the "charisma" attached to these areas is well deserved. Each year, American industry expends more energies in research, marketing, manufacturing, advertising, and other areas in order to run the race for new products. Never before in the history of American business has it been so true that "the race belongs to the swift."

The objective of this chapter is to present an overview of new product development. Few people today question the need for new products, and this chapter expounds upon those needs, both for industrial and for consumer products. The belief that New Product Development can be planned, programmed, and systematized to a high degree is manifested in the New Product Model for consumer products and the evolutionary steps through which a new industrial product proceeds. The subject of why new products fail is touched upon, and ten philosophies of successful New Product Development are listed. The role of research, sometimes underrated in these days of market orientation, in developing new products is presented from the research point of view.

For the reader who has sufficient interest, several case studies of the "WISTU PAPER COMPANY" are provided, with their proposed solutions. This chapter will take on more meaning if read in conjunction with Chapter 2, Marketing Research In the Specialty Paper Business, and Chapter 6, The Use of the Computer in New Product Development.

A look at the record shows that the paper industry does not have to take a back seat to any in the new product development area. During the period from 1952 to 1966, it has been estimated that in consumer paper products, the number of items in a typical supermarket had increased 179%, that is from 52 to 145. The impact of nonwovens, in consumer, institutional, and industrial markets, is just now entering an era of dramatic growth. Carbonless impression papers for the business forms industry, coated office copy papers, and plastic coated printing papers are all examples of products widely used today that were not well known several years ago.

If there are winds of caution that should be watched by new products people in the paper industry as they chart their respective courses over the next several years, they are probably in recognition of the fact that some important new paper products will probably be developed by companies closely related to, but not directly a part of, this industry, as these companies utilize the systems approach more and more.

II. THE OPPORTUNITIES NEW PRODUCTS CREATE

We might ask ourselves why so many American companies devote so much time, money, and effort to developing new products.

Ironically, the very same things that cause some people to be "too busy" to develop new products, e.g., competitive pressures, declining markets, price attrition, and other negative factors affecting existing businesses, also cause New Product Development to be a major activity in most progressive companies.

New Products, then, are necessary for several reasons:
1. To replace those products or product lines that are becoming obsolete because of competitive pressures
2. To enable the corporation to meet its long-range planning objectives in the areas of sales and growth
3. For diversification purposes
4. For development of "spill-over" technology
5. To enable the company to penetrate new markets with existing products.

1. To Replace Products That Are Becoming Obsolete.

Defining that point in time at which a product becomes obsolete can be a most difficult, if not impossible, task. For example, the hula hoop

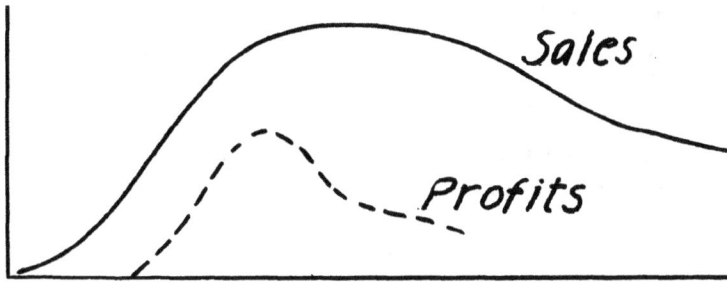

Fig. 1-1. Life span of product.

may arrive on the scene in 1955, be obsolete in 1958, and have a resurgence in 1967, while a keg of old wine may become more and more valuable with age. Generally speaking though, products, like people, have a life span, during which they are born, live and die.[1] This can be depicted graphically.

Fig. 1-1 shows that sales come before profits, that profits are highest early, and later sales level off, or even drop, while profits become more and more marginal. This "erosion" in the life span of most products occurs because *economically speaking* users (customers) place a higher value (price) on some other product that performs an analogous function ($100 horses sell for $50 because most buyers place a higher value on the automobile as a form of transportation.) One concept, which is exciting here, is the thought that as a New Product endeavor, we might choose to make obsolete our own products, thereby beating our competition to the punch.

2. To Help Meet Corporate Sales And Growth Objectives

Most major companies in America today have a Long Range Planning Function. Part of that function is to develop corporate growth objectives. These objectives, when stated, often take the form of—"The ABC Corporation will strive to achieve sales of X dollars and profits of Y dollars in year Z." It is apparent that this objective can be realistically achieved in only a few ways, among them: (a) increased unit sales of existing products, (b) increasing prices and/or decreasing costs (which may or may not meet sales and growth objectives), (c) participating in acquisitions or mergers, and (d) development of new products, either internally or externally. Here, we are discussing (d).

To recognize the impact on corporate growth that new products

---

[1]The concept of a product life cycle is treated more fully in some marketing texts.

can have, we should know that many companies, which today have annual sales of hundreds of millions of dollars, have over 50 per cent of those dollars coming from products that were not commercially available ten years ago!

Furthermore, even if a company had *no* growth objectives, because of erosion of sales and profits depicted earlier in the product life cycle concept, a steady influx of new products would be needed just to maintain existing levels of business.

3. For Diversification Purposes

The development of new products frequently places a company in new markets, which it never before served. This can happen by design, of course, as when the product was planned as an entree into a new market; but sometimes it can happen by chance, or when a customer, without the manufacturer's knowledge, figures out a new end use for the product.

4. For "Spill-Over" Technologies

New technological know-how obtained in the development of the new product often can, with some imagination, be used to upgrade existing product lines. This can "slow down" the erosion of the product life cycle, or even be a springboard to an entire family of new products. This can happen even when the New Product fails—for example, if a technology or even a business relationship with another company is exploited. An example of this could be where a paper manufacturer and a typewriter manufacturer attempted unsuccessfully to develop a *better* typing paper, but as a result of the right people coming together at the right time, a "better" adding machine tape was developed.

5. To Carry Existing Products Into New Markets

It can happen fairly frequently that the excitement with which a successful new product is received in a new market area creates a positive company image, on the strength of which more established products can ride, but where they could not penetrate that market on their own qualifications. This could be considered a form of market synergism.

III. THE PROBLEMS NEW PRODUCTS CREATE

Many people have written many things in the last few years about new products.* Hardly a day goes by that we are not presented with

* See Bibliography.

either through magazines, newspapers, television, or radio, a "new this or new that."

Despite this proliferation of new products, however, there is much about them that we do not understand. If we did, the batting average of successfully developed new products would be much higher than it is. It is still a hard fact of business life that most new product ideas fail.

In addition to the fact that most new product ideas fail, in many companies, their development also causes stress and strain on existing corporate businesses. The reasons for this become quickly evident when we think about them for a moment.

In today's highly competitive, results-oriented business environment, corporations are committed to the daily struggle of maintaining or increasing their share of the market, of keeping costs as low as possible, of coordinating advertising and promotional programs with manufacturing schedules, and in general, going about the daily task of keeping the business running. Because of this, New Product Development should be a staff function, and should *not* be the responsibility of those people who have a daily manufacturing or line sales quota to achieve.

Although organizational responsibilities of people are the most sensitive areas of disagreement upon which a new product can impinge, these are not the only ones. Equipment scheduling, especially during peak business periods, for new product trials vs. existing product runs for customers can become trouble spots. The customer normally takes priority, and this is as it should be. The new product, however, must be given its fair share of time on the equipment as soon as is reasonably possible, or else the project will never be completed.

Finally, the interface between market research and development and technical research and development can become a real "demilitarized zone" of buck passing if the proper atmosphere of cooperation is not created. These "two sides of the same coin" must work in harmony, and as a team, if successful new products are to be achieved.

IV. THE NEW PRODUCT DEVELOPMENT CYCLE[2] FOR INDUSTRIAL PRODUCTS

A. Search For New Products

If we accept the concept that new products are desirable, and even

---

[2]The separate phases of New Product Development can be described in various ways. A very readable version is Booz-Allen and Hamilton Incorporated's *Management of New Products*.

necessary, to the business firm, our concern becomes one of knowing where to go for new product ideas.

For a search to be properly organized, the business must be properly defined. This is not always easy.

For example, is a manufacturer of stereo-record players and television sets in (1) the large appliance business, (2) the entertainment business, or (3) the electronics business? If (1), a new product consideration might be a refrigerator line; if (2), the acquisition of a baseball franchise might make sense; if (3), development of a military contract might be their objective.

To the uninitiated, it may come as a surprise that there is no lack of new product ideas quantitatively. What is lacking is a large number of quality new product ideas. This brings us, then, to the first step in a formalized sequence of events that occur in the commercialization of a new product idea. This first step we can call SEARCH.

The search for new products proceeds in many directions. Salesmen, housewives, suppliers, company files, trade magazines, patent literature, and newspapers can all be sources, to say nothing of inventors, research and engineering studies, universities and colleges, and government agencies. Although new product ideas do come from a multiplicity of sources, most of these sources fall under one of these four categories: (1) Modifications of existing products, (2) Market research studies, (3) Research and engineering technological breakthroughs, (4) "Blue Sky" or "No holds barred" creativity projects, sometimes referred to as *Ivory Tower* thinking.

1. *Modification of an Existing Product*

In many companies, the New Product Staff Group does not concern itself with these "product modifications." They are considered to be in the province of the existing marketing managers, who, with their daily contacts through their field sales force, are better qualified to react to the changing needs of the marketplace through the vehicle of design changes in existing products. This does not mean, of course, that such new products are not a very necessary part of the maintaining of the company's sales and profit levels.

2. *Market Research*

The most highly probable area of success in developing new products comes from being market oriented. This simply means going to the customer, discussing his operation, processes, problems and needs, and designing and developing products to solve these problems and fulfill these needs.

3. Research and Engineering Technological Breakthrough

When we recognize that in the history of the world most of the physical scientists that ever earned their doctoral degrees are alive today, we begin to get an appreciation of the technology explosion we are experiencing.

Many wonderful new things are invented in research, and frequently no one knows exactly what these things are good for. This is an area where creativity becomes critical. The ability to see things differently is required. Good ideas should not be locked in a dark closet until "tomorrow", because tomorrow never comes. They should be exposed to the sunlight of the market place so they can grow.

We don't always have a Galileo, or an Edison, in our midst. But penicillin was discovered 50 years before Alexander Fleming observed the bacteria-devouring mold. The difference was that Fleming recognized he had something. When Charles Goodyear accidentally vulcanized rubber, he knew he had come upon a real opportunity. Conversely, Edison tried many materials before developing the electric light, whereas Carruthers worked many years to develop nylon, without being sure of what kind of a market was out there.

4. "Blue Sky" or "No Holds Barred" Creativity

There is a certain type of creative thinking that occurs when the environment is such that the submitters of the ideas know the ideas are not going to be viewed judicially as soon as they express them. It is from this this free surging mental gymnasticism that tomorrow's great new businesses will come. Pick an area for openers—Oceanography —Pollution Control—World Hunger—put a small group of bright people of mixed backgrounds in a room for a week and see what they develop.

Conceivably, such a group could have come up with a photocopying process technology. A student could have expressed a problem—the laborious copying of notes—a photographer may have suggested taking a picture—an engineer might have pointed out the desirability of a less demanding process—a physical chemist might have suggested the use of photoconductors or thermal-sensitive coatings.

The search for new industrial products, then, consists of new product ideas, both inside and outside your own company and country, and formalizing a procedure through which these concepts can be commercialized into successful new industrial products.

B. Preliminary Screening

After completing the search for new product ideas, a preliminary screening takes place. It is at this stage that the following questions are asked:
1. How great is the need for this product?
2. How feasible does it appear to be to develop it?
3. Does it seem roughly that the costs would be in line?
4. How does it stack up against those products against which it would be competing?
5. How does it fit the company's (1) manufacturing capabilities, (2) marketing capabilities, (3) distribution channels, (4) existing product lines?
6. How does it fit the objectives of the corporation?
7. How does it fit the weighted priority model as a tool for screening, if we use such a tool?
8. What is the present business climate, and how will it affect this proposal?

C. "Proving" Technical Feasibility

If the product idea successfully passes through the preliminary screen, it is necessary to know how feasible the concept is. This does not mean that absolute "proof" is required that the product can be made, but rather than a reasonable chance of success exists to make the product. This is a good time to get the "experts" together in the form of a screening committee. Representatives of research, engineering, manufacturing, marketing, planning and new product development should be involved here. If possible, a prototype, e.g., a hand sheet, or a trial roll, should be made in the lab and tested for functional properties.

D. Determining Market Acceptance

The validity of the product concept, and the prototype, if available, should be tested in a controlled market area. When industrial marketing is involved, twenty customers may represent the national market, and initial acceptance is relatively easy to determine. If consumer products are involved, a very sophisticated market test may have to be designed.

E. Business Analysis

This step in the product evaluation process is extremely critical, because beyond this point, the investment to finish the development tends to be large. The business analysis should cover the following points:
1. Size of market
2. Market share
3. Market growth trends
4. Estimated selling price
5. Estimated manufacturing cost
6. Estimated other costs
7. Estimated profitability

The business analysis can be performed in several ways. One of the most efficient ways is by use of the computer. (See Chapter 6).

In estimating size of market and market share, one must be aware of segmentations within the market. A given product line may be required where modifications of the basic product, either for functional or economic reasons, may be necessary for the new product to penetrate the market properly.

F. Product Testing

If technical feasibility, market acceptance, and profitability all signal "go," the next step in the sequence is to make some product and obtain trials in the market. Again, we must differentiate between industrial and consumer marketing. In consumer marketing, these trials would most likely take the form of a market panel test. In industrial marketing, they would probably involve trials in the customer's plant. Frequently such trials indicate areas where the product performance is deficient. Thus, the result of the trial is a recycling of product redesign, prototype manufacture, and retesting in the lab and the field. PERT[3] diagrams are useful here as a planning and control device.

When it is believed as a result of such testing that the product design is close to being finalized, a scale-up to manufacturing equipment is necessary. If this is "new" equipment, typical start-up problems can be expected. Even on existing equipment, new products present a myriad of problems until the operators have learned to run it. Initial runs can be expected to vary widely in quality.

---

[3]Program Evaluation and Review Technique.

G. Marketing

When quality control is within tolerances, product acceptance verified, and profitabilities found to be within target based on latest calculations, sales are sought for the new product.

Such sales do not ordinarily come easily, nor, in the beginning, are they very large. Customers that are willing to pioneer new developments must be found. Frequently, these initial customers are the same ones that did the trial work. Frequently they are *not* the largest concerns in their particular fields.[4]

As sales begin to grow and product acceptance in the market place is verified by the receipt of repeat orders, a turnover of the new product to a marketing group, where one exists, is initiated; and where none exists, its beginnings are established. In either case, the "turnover" should be well planned and well timed.

Product orientation seminars for the sales force may be necessary. The new product project manager may be temporarily or permanently assigned to the marketing group. No matter how it is done, the important thing is that it is done in such a manner that the product is not harmed during its infancy because of a faulty turnover.

Pricing schedules, terms, policies, inventory levels, packaging, labelling, advertising and promotion, all have to be phased together for the new product to get a healthy start in life.

V. CONSUMER NEW PRODUCT EVALUATION*

A. The Need for New Product Development

The necessity for efficient systems for new consumer product introduction is generally widely recognized. Every year, many hundreds of millions of dollars are spent on new consumer products. Only a small proportion of these new consumer products ever reach the market and only a tiny proportion of these are financially successful. For example, in the paper industry's primary consumer products distributional field, that is in grocery products, it has been estimated by This Week Magazine that about 2,500 new food and grocery products are

---

*The thoughts expressed in this section were originally presented on 4/21/66, to AMA Seminar #5240-12, New York, N.Y., by Mr. Gilbert Dimentis, Director, Consumer Marketing Research, Kimberly-Clark Corporation.

[4] See Raymond E. Corey, *The Development of Markets for New Materials*, bibliography.

presented each year to the buying offices of chain food stores. Of the 2,500, about 1,300 are accepted for distribution by the chains, while 1,050 old items are dropped from the shelves.

Consumer paper products have shown one of the fastest growth rates in item counts. As mentioned it had been estimated that in this field, the number of items in a typical supermarket had increased 179%, (from 52 to 145) from 1952 to 1966. Although the percentage increase is large and important, the average number of net additions in the paper field has been only about seven items per year.

In this perspective, the operational question then becomes to handle and screen a large number of new ideas that are being continually generated, each with fairly low probability of commercial success, or to deal with a much smaller number of final products, which have high probabilities of success. Experience indicates that this screening process requires formal programming and a discipline that had not been necessary in earlier times. The use of a system such as the New Products Model to be presented is one effort to inject this necessary discipline into a system form.

B. Applicability of the Model

The New Product Model presented is one that especially pertains to a mass distribution-consumer product operation. This involves the marketing of high turnover branded products, which are sold through grocery, drug, discount and variety-department stores.

The customers are primarily female homemakers. In addition to these qualifications the new products also generally have the following marketing-financial characteristics:

1. A high capital investment in marketing inputs. (It is not uncommon to invest five million or more in capital marketing dollars for a new product introduction.)
2. A relatively long product life cycle
3. A relatively high level of research and engineering costs to develop both product and production facilities
4. A relatively high dollar and time investment in creating production capabilities

Because of these characteristics, it is highly desirable to screen down the large number of new product ideas with low probability of success into a much smaller group of ideas with a relatively high probability of success early in the product development cycle. It is *highly* desirable to do this screening before research or engineering development takes place or any traditional test marketing is programmed.

For those firms having different sets of production or marketing requirements, a plan such as the one being presented may be too elaborate,
too time consuming, or plainly too expensive. However, the Model
being presented does have wide aplication and, with appropriate modifications, should be helpful to those concerned with the industry.

C. The New Product Model

Let us now turn to the flow diagram outlining the steps in the New
Product Model. Although this model may look complicated, such a
procedure is extremely useful as a planning tool.

The first step in the Model is labeled "Corporate/Division Product/
Profit Criteria." This is the Strategic Corporate Plan or the general
Franchise Determination step.

D. The Strategic Corporate Plan-Franchise Determination

Preceding any generation of specific ideas or concepts must lie a
strategic corporate plan or a general franchise determination. This
is the overall frame within which new and existing products must "fit."
*This requires that the corporation (or division) decide what fields of endea-
vor are appropriate (or inappropriate) for its present and future operations.* The appropriate fields for present and future operations are
usually determined by the structure of the company resources and its
anticipated applications to present and future problems and opportunities. For example, an operating area may be determined by one or more
of the following considerations:

a. Present distribution systems
b. Present marketing capabilities
c. Present brand franchises
d. Present technological skill
e. Present productive capabilities or knowledge
f. Capital availability.

Also, a strategic corporate plan would contain certain financial
limitations such as return on assets and return on sales.

The critical definitional need here is for an *individual, unique* marketing
concept.

For example, the convenience food manufacturers are not marketing
just particular kinds of food or calories but are selling time savings to
the housewife (or in some cases a sense of accomplishment, or an anticipated compliment from husband or children, or "fun", or some other

intangible psychic value).[5]

The determination of an appropriate and workable corporate franchise definition is the first and most important screening step in any New Product process. This definition critically determines what any company can and can't do and what it should and shouldn't do as a marketer.

1. *All Sources of Information*

Assuming that the first problem of corporate franchise determination is successfully solved, the initial step in this programmed New Product system is the collection of data pertaining to perceived and latent consumer needs and wants. This collection of data most efficiently *precedes* any new product concept or idea formulation for one simple reason. It has been found to be substantially more desirable and efficient (less costly) to design solutions (new products) around real consumer problems (needs/wants) rather than to haphazardly design solutions and then see *if* a problem exists.

A myriad of sources are available that will grossly indicate whether consumer wants/needs exist or will exist in certain areas of such consumer activities. Some of these sources are:

a. Consumer complaint or suggestion letters
b. Trends in general behavior patterns (changes in the sociology of the community at large such as more leisure time, fewer servants, more two-car families, earlier marriage ages, and more in-home washers.)
c. Programmed need searches utilizing techniques such as:
   1. Behavioristic studies of individuals in certain environments (how do women *act* when they are washing dishes, washing windows, and cooking?)
   2. Group sessions or individual interviews on attitudes toward, likes and dislikes, or problems, or satisfactions, or lack of satisfactions in certain kinds of activities.

Of course, there are other sources such as research and development (R & D) laboratories, suppliers, and competition that may also suggest opportunities for better servicing consumer needs. The principal thing to remember is that a new product is not something that should be defined in physical or engineering terms but a set of consumer needs or desires the fulfillment of which can be facilitated by some physical reality that we call a product.

---

[5]Theodore Levitt, "Marketing Myopia," *Product Strategy and Management*, pp. 8-26.

2. *New Product Concept/Idea Formulation*

After the determination of an appropriate "set" of consumer wants/ needs, the next step in our New Products Model is the activity that embodies these needs into a product concept that relates needs to the attributes of a product. The key to this step is to relate *perceived* consumer needs to *perceived* product attributes (*Perceived* need satisfaction) and *not* necessarily to real needs or to real attributes. The investigations or thinking in this area should primarily be "listening" to the consumer oriented rather than "talking" to the consumer oriented.

Appropriate members of a New Product Concept team should include individuals such as the New Business Manager, the communications experts from both the company and the advertising agency, the market researchers, and representatives from R & D and manufacturing. The objective here is to create the "bare bones" of a product in word form to meet the earlier developed needs set of the consumer.

Once a product concept or series of concepts is created in a form that enables the consumer to see the *relationship* between the perceived attributes of the product and her perceived needs, it is then tested against the consumer to see if the linkage has been successful.

3. *Rough Concept Screen*

The methods of concept testing can run from the presentation of simple statements of attributes to more elaborate semifinished advertising forms. The presentations to the consumer can take place in the form of personal individual or group interviews or by mail, depending on the problem.

For several reasons, it is more efficient first to rough-screen concepts. This allows for less expensive screening of the large number of concepts available at this stage and also to gain broad information as to future market segmentation possibilities.

The rough screening process divides our original group of concepts into three separate piles: those indicating a "no-go" situation (which are tossed into a reject pile), a "go-qualified" (which indicates further refinement or redefinition of concept is necessary), and a "go" situation (which is processed to the next step in the Model.)

One further word about concept screening would seem to be in order here. It is that concepts are usually, if not always, measured on a relative rather than an absolute scale of acceptability. This means that a number of concepts must be screened, utilizing the same system of measurement. There is probably no substitute for a certain amount of experience in what the various indices of "willing to try" or "willing to buy" or "interest in" mean in terms of predictable consumer purchase behavior.

4. *Technical and Economic Feasibility*

Once a concept has passed the first rough screen for acceptability, it is usually subjected to a technical and economic feasibility review. This is an order-of-magnitude investigation that would indicate the general probabilities that the projected product attributes could be emobodied in a product, that the product could, in fact, be manufactured, a *rough estimate* of the probable cost of manufacture, a projected manufacturer and retail price, a rough estimate of probable sales volumes, marketing costs, and profitability.

The purpose of this stage is to determine the appropriate levels of some of the other consumer attributes of the product (such as retail price) to be further examined and also to determine the limits of the production—marketing area in which we might be operating. See Chapter 6, Use of the Computer in New Product Development.

5. *Concept Test/Structuring and Advertising Development*

At this stage of development the "bare bones" concept starts to take on flesh in terms of such marketing attributes as branding, size, color, price, odor, and packaging. In other words, the communication to the consumer begins to get more explicit in terms of the various "cues" that are available in the real world marketing situation. Although the verbiage may still be in a simple manner, the concept and its communication are now nearer final forms.

The marketing and communications experts play a key role at this stage. The consistency of the "brand image" is now being created. The structuring and positioning of the product is done at this point. All this must be consistent with the perceived needs of the consumer!

The now more complete package of attributes are again presented to consumers for their review. Does it still meet their perceived needs? Are there segmentation possibilities? What should be the target audience? Again, the screen operates to reject, recycle, or allow the refined concept to pass.

6-7. *Advertising Development and Testing*

Again, assuming the concept has received a "go", it proceeds into the advertising development and testing stages. These stages imply that product attributes, no matter how good, must be communicated to be effective. The creation of this communication is considered to be critical. It must not only be effective but also be a reinforcing element in the creation of the final brand image and in again linking the perceived attributes of the accepted concept to the perceived needs of the target audience.

The media needs must also be considered here in terms of realities of size of market, structure of market, and target audience.

The result of the advertising development stage is usually a semi-finished T-V commercial or print advertisement, complete with full copy platform.

The finished communication is again tested against consumers to see if it is effective in terms of predetermined "norms." The techniques for this testing are many—such as:

1. In theater, in mobile van
2. In home
3. On the air
4. Split runs.

Again, the concept is allowed to proceed, to be recycled or rejected. The "go's" now move into Step 8.

8. *Research and Engineering Development*

Research and engineering development starts to take place in Step 8. Up to this point only *abstract* items such as ideas, concepts, consumer wants, and communications have been reviewed. This is the first time "things" are involved in our operations. At this point the scientists and engineers *begin* their contribution.

It is now possible for these highly talented (and expensive) people to create and design in an extremely effective manner since the product attributes and perceived needs are now clearly spelled out (and evaluated). There is relatively little danger of undisciplined wanderings of a research project if market needs are correctly communicated to the scientist. However, there is still plenty of room for the creative research man to bring an abstract perceived need into the reality of a three-dimensional product. There is also the imputed satisfaction to the R & D man of knowing that the product of his mind and effort has a relatively high probability of reaching the "real world" and that he is not merely "spinning his wheels."

At the completion of this stage not only must the product be created to meet consumer perceived needs, but it must meet mass manufacturing and cost and profit requirements.

9. *Product Prototype*

Assuming these can be accomplished, a product prototype or a series of prototypes is created (Step 9). This is usually done without large capital expenditures (but not always). At this "hardware" stage the product development costs start to rise rapidly. Assuming an adequate number of prototype units can be produced, we move into Step 10.

10, 11, *and* 12. *Limited or Extended Use Testing with Advertising*

Our next step is to take the prototype units with the previously tested advertising into limited or extended use tests. This is to determine if the product delivers the benefits as previously perceived and tested and if it satisfies the consumers' needs as previously perceived and determined. In fact, if properly structured, this stage can be thought of as being a small scale market test.

Depending on the results of this stage, it may be decided that a large scale market test is unnecessary. Many companies, where the capital equipment and/or marketing risks are low or nonexistent, proceed from the extended use test into national introductions of the product.[6]

This in-home testing of the product again yields additional information on how the consumer is likely to perceive the product under more "real life" conditions. Is it seen as a luxury product or an "everyday" usage item? Is it for young people or for old? Is it to be used for a single purpose or in a multiplicity of situations? Does the product have attribute flaws? Could it be improved by a change of color, of scent, of packaging, of usable units within a package?

In addition to the previous kinds of questions, this kind of research frequently uncovers *unexpected* product virtues that can later be utilized in an advertising or promotional campaign. At this point in our screening operation, we are collecting the final data, on a very *specific* basis, on target audiences, advertising platforms and executions, product attributes, and pricing.

The test techniques utilize in-home use, either on an individual unit (Monadic) or comparison basis. The length of test is determined by some compromise between the necessity for moving forward (and lower

---

[6]Recently (1968) some consumer goods companies are market testing advertising of the product but not the product per se in an attempt to speed up the new product cycle.

cost) and the desirability of determining what the real world "repeat" and usage rates are likely to be. For example, we want to know if the product is a "novelty" product, which the consumer will purchase only one or two or three times. Will the product be used once a day, once a week, or once a month? How many ounces, sheets, or servings will constitute the average use? Is the product used in a specialized context or in a number of situations? What is it competing with? Our established brands? Competition?

Depending on the results of the use tests, the product can be recycled to the Research Department for further development or recycled for advertising revision or brought forward to market test or regional or national introduction.

If the full "go" situation was reached in Step 10, we now have a new product with perceived attributes that meets a perceived need of consumers. The projected communication system relates these two perceptions in an effective manner and the product lives up to its communication promise. The target audience has been defined and its use behavior in reference to the product has been established.

Assuming we are still in the "go" situation, we could move all the way into Step 13—The Market Introduction Decision.

13. *The Market Introduction*

The final step of our New Products Model is the Market Introduction phase. This will involve the now completed product-communication entity into one or more "real world" markets. The objective here is to test, and refine, if necessary, the results of all our earlier efforts. Running audits of consumer perception of communication, of initial purchase, of repurchase, and of distribution, are usually made. Alternatives of communication levels and of consumer initial purchase incentives, are usually introduced at this point so that the most economically effective national program can be developed.

The large dollar expenditures in testing are made here. Large production runs of product, large media expenditures, large selling costs and large market research costs are involved. A small city test market operation usually costs in excess of $100,000 and several are usually needed. Prescreening costs before this point are usually less than this, and for this cost many, many concepts have been screened out, redefined, and improved.

Conclusion

In summary—

1. It's the perceived needs of people that come before the concept (the perceived solution of the needs problem) or before the product.
2. The concept, that is the perceived problem solution, must fit the perceived needs set.
3. The communications must relate the problem to the solution of the problem.
4. The product must deliver the problem solution promised in the communications.

In spite of whatever is done to program and screen new products, we do continue to live in a probabilistic world and not all new products are successful. One fact stands out, however, and it is that proper programming and discipline help to increase the economic chance of success and reduce the economic failure rate. As in the past, the secret of continuing success in the new product field is to keep the mistakes small and to run hard with the successes.

VI. THE ROLE OF TECHNICAL RESEARCH AND DEVELOPMENT IN NEW PRODUCT DEVELOPMENT

The prime role of Technical Research and Development is to invent, innovate or adapt. In most cases, therefore, Research and Development activity is necessary to the New Product venture. The needs of the marketplace are first identified and categorized, and the fulfillment of those needs must subsequently be built into a specific product produced by a workable process to provide a profit. Often R & D serves to provide a bridge between needs and satisfaction of the needs.

Technical Research

The major end product of Research is information. This often consists of a detailed description of the composition and form of the new product and approximately how it should be made. Once this information has been acquired and turned over to the Development group, the Research activity can probably be reduced and constitute a consultation service to other groups.

Industrial Products

Typically, the broad market needs are uncovered by Marketing or Market Research personnel. If invention is required to design a suitable new product, Research help is requested. A working team is often formed to interact directly with technical people of potential customers. The objective of this Marketing and Research team is often to translate the broad marketing opportunities into specific technical requirements. The more the teams can learn about the intended end use and the customer's capabilities and own processes, the more effective the research program. At this point the researcher hopefully begins to fulfill his true role as a creative person. The program during this period is technically oriented rather than market oriented. The Researcher draws on experience and knowledge to design the new product. No longer are words and specifications sufficient. An actual sample of a product is now necessary; available materials must be combined on a laboratory bench basis to provide potential customers with small amounts of test material. Hypotheses are set up, the feasibility examined experimentally and the resulting samples evaluated either in the laboratory or by the customer. The product is modified, and if necessary, redesigned. The product design may be made by methodical approach, or it may come from a "flash of genius." The invention may be easy or it may be difficult, but it is invariably hard to schedule. As the product is continuously evaluated and improved, the customer's needs become better defined and the specifications become more exacting. In due course, the composition and form are defined, the approximate method of manufacture is determined and the orientation is now toward the Development phase.

Development

The role of Development is to reduce the laboratory design to practical manufacture. It must be determined *exactly*, not approximately, how the new product is to be made. The development team must work with Engineering personnel in designing the commerical process. They must work out problems of manufacture with the Production and Quality Control people during the shake-down period. It is necessary to work closely with the Marketing group to provide commerical trial quantities of material for customers and field evaluation even before commercial equipment is modified or available. In short, the Development group works with either pilot or full-scale manufacturing equipment in order to translate the information supplied by Research into a product with commercial reality. At this stage PERT or Critical Path techniques are effectively used as an aid to realistic project scheduling.

Consumer Reports

The task of R & D in the Consumer Product area is in many ways similar to that required for Industrial Type Products. The major difference is in the degree of customer relationship. The interaction and playback is generally not with a specific customer or even a limited group of customers, but it must somehow be with the general public. All the talents of the Market Researchers delineated in Chapter 2 must be brought to bear on the problem. Small samples rarely supply adequate playback and recycling, so considerably more and elaborate sample production is required. Development and Engineering are involved at an early part in the project to produce the greater amounts of material needed to prove the product to a large enough market population.

The objectives are the same, however, in each area. The conceptual design must be translated into a feasible test product, the marketplace must accept and change the product to fit its real needs, the exact manufacturing process must be designed, equipment must be built and proved, and finally product manufacturing must begin.

Personnel

There must be an interplay of technical and marketing personnel or capabilities to effect a smooth transition to and through the various phases of a New Product Development project. The principal orientation at any given moment may be Marketing, Research, Development or Engineering or some combination of these functions, before the project is satisfactorily turned over to Production and Sales. Representatives of each of these areas should always be involved to supply specific data or support when needed, to maintain background for their period of principal orientation, and to function as consultants to the project subsequent to their active role.

In a small, simple company or project, one person could conceivably perform all the functions. At the other end of the spectrum with a large company or a sophisticated project, large groups of men could be involved in each phase. The literature has identified some of the problems of organization and communication inherent in these large endeavors. Russell Peterson of DuPont and Robert Wolfe of B.F. Goodrich have shown how their companies organize to bring New Products through the gestation period and into the marketplace.[7] It is important to recognize what problems could occur in any given organization and to cope with them in advance.

[7]Russell W. Peterson, "New Venture Management in a Large Company," *Harvard Business Review*, May-June, 1967.

VII. WHY NEW PRODUCTS FAIL

It is sometimes tempting for a company to rationalize the lack of success in new product development by believing they are part of an industry that is noninnovative in nature, whereas other industries are more "progressive." The facts are that, like most business successes, successful new products are the result of good management, and the industry has little to do with it.

Because many new products fail, much thought has been given to the reasons for these failures. Some of these reasons are discussed next in this chapter.

If the New Product Development cycle is so formalized and straightforward, one might ask why so many new products fail. There are several reasons.

A. Length of the cycle

The average new product in American Industry can take three to seven years to develop. This is a relatively long time. Since the market place is dynamic, and not static, many things change, including the data upon which the original new product estimates were made.

B. Inadequate market research

This is probably the most significant single cause of new product failure. Without a "good feeling" for what the marketplace is seeking, it is extremely difficult, if not impossible, to design the product properly. Erroneous market data can arise from a poorly planned questioning structure, or from a poor interpretation of the data, or from several things in between, such as the respondents not being a representative sample of the industry, or asking the question at the wrong point in the distribution channel.

C. Faulty Business Analysis

It is difficult in the world of business to keep costs of existing products under control. When dealing with new products, where no historical cost standards are available, where new equipment or little-tried processes are involved, plus new distribution channels with their attendant marketing costs, it is much tougher. Thus, it is not infrequent for the new product marketing manager to learn that costs are 10 per cent higher than originally estimated, while selling prices are 10 per cent lower, and profit targets go out the window.

D. Competitors' Activities

Competitors are pretty smart people. Sometimes just as we are about to introduce the new product, one of them introduces one very much like it, or the competitive product to which the new product was so superior is suddenly upgraded, or its price is reduced, or a heavy advertising program is initiated for it. Any or all of these actions can create a new situation.

E. Insuffcient support

New products are like human babies. They need extra special care until they are strong enough to stand on their own two feet. Frequently, however, such extra support is extra expensive, and, therefore, not given. As a result, some good new products never grow up to become successful. Yet in the same company, hundreds of thousands of dollars may be being spent to nurture along an octogenarian as it totters to the grave. New products fail, for other reasons but these are some of the main ones.

VIII. TEN PHILOSOPHIES OF SUCCESSFUL NEW PRODUCT DEVELOPMENT

In review, then, we have seen why new products are desirable, and even necessary. The search for new product ideas was discussed, and a stepwise procedure for successfully developing new products was outlined. Finally, some of the reasons for new product failure were mentioned.

At the end of this chapter several case studies of new product development have been included. The reader with sufficient interest to "play the game" will also find the solutions to the problems of the "WISTU Paper Company" provided.[8]

In closing, some of the "philosophies" of New Product Development are listed for the reader's benefit.

1. New Product Development must be supported by top management.
2. New Product Development should be a staff, not line, responsibility.

---

[8]These cases were originally written in the spring of 1968 by Mr. Sacher for the use of student members of the Society for the Advancement of Management at Wisconsin State University, Oshkosh, Wisconsin, during a seminar it sponsored on new product development.

3. New Product Development should be formalized, planned, controlled, and managed like any business function.
4. The members of the new product team should be "volunteers", highly motivated to run the risk, and be willing to be compensated according to the success, or lack of it, of their track record.
5. The New Product Development Group should not be a profit center, but should be viewed as an area where dollars are invested today for return tomorrow.
6. The New Product effort should be performed with sufficient authority that it receives its fair share of research, production, and marketing time, and it should not always be a lower priority user of these corporate resources than are the existing grade lines.
7. Each new product project should have its own budget. Cost accounting should reflect all expenses incurred against it, and all revenues generated for it, until it is turned over to the proper marketing group.
8. The interface between research and marketing in product development work should be such that each group is consulted by the others for thoughts and opinions in all areas.
9. The new product must be adequately supported by the corporation.
10. The successful project team should be given the option of moving out with the product, or of beginning over again on other new projects being started in the New Product Development Department.

IX. CASE STUDIES
CORPORATE POLICIES—WISTU PAPER COMPANY
May 18, 1968

Your Board of Directors combines an aggressive marketing policy with a conservative investment policy. Thus, it strongly encourages development of those new products that "fit" existing manufacturing capabilities. Although it judges any new product proposal on its own merits, your past experience with the Board has made it clear to you that it looks for a higher return on investment for those projects that deviate from existing know-how, than for those that "fit".

All new products must generate a minimum return of 6% on sales, and 12% on investment.

"WISTU" is a leader in the industry and does not usually "Overreact" to competitive pressures. The company has a long history of having introduced new products that are really unique, not just "me-too" products. Sales gives great credit for this to the fine research staff at WISTU, which has never been known to say "we can't do it."

When money rates exceed 5-3/4% for prime borrowers, WISTU generates funds by stock issues.

All departments are required to follow all financial department policies at all times. Manufacturing costs include depreciation expense for existing and proposed new equipment.

Data in this case study are disguised.

To: Marketing Manager
SUBJECT: BOOK PUBLISHING PAPERS
Date: May, 1968

> Presently, five paper companies compete for a major share of this business which amounted to 1,000,000 tons last year. For the past ten years, the market for books has increased 15% a year. Conservatively, you believe you could capture 10% of this market in the first year, 15% in the second, and that this share would level out after reaching 20% the third year.

> The market is sold through distributors, and is highly concentrated in the East. Your mills and your merchant representatives are well located to serve this Eastern market. Sales are 50% in rolls, 50% in sheets.

> Sales has assured you they can effectively compete in this market at a $250/ton selling price, freight prepaid. Freight charges are $20/ton. Most of your present grades sell for $200/ton, F.O.B. mill (customer pays freight). Selling expense on a grade of paper this sophisticated will be 10% of gross sales.

To: Manufacturing Manager
SUBJECT: BOOK PUBLISHING PAPERS
Date: May, 1968

> You have the machines, technology, and process control

to be able to manufacture the main sheet and roll grades at competitive costs ($150/ton).

Producing a nonglossy finish paper that prints well is relatively new, and equipment to produce such results is not commercially available. Estimates are that a full-scale unit would cost $200,000 to $400,000. Other machine investment is charged off at $100/ton.

The manufacturing department has learned from purchasing that the material to add opacity to your paper has almost doubled in cost, and the cost of manufacturing book publishing grades is now $8 above their present grades.

BOOK PUBLISHING PAPERS
R & E

Currently 90% of book publishing paper is glossy to provide a good printing surface. A competitive company has recently introduced a nonglossy finish that greatly improves readability of this paper. Sales reports that this sheet is being well received.

Several trials you have made indicate your paper is well accepted by the publishers although the price you can command is now $240/ton, not $250.

Engineering estimates they can modify existing equipment to produce a nonglossy finish in 18 months for $100,000. They can do the job in six months for $200,000. This is a project the marketing manager gives his full support to, even though research believes there is only a 50% chance of successfully producing a nongloss finish this way.

BOOK PUBLISHING PAPERS
Planning

Approximately 100 tons/day of your current sales are in a marginal profit line because of competitive pressures. Forecasts indicate that over the next three to four years this will reach 300 tons/day.

By 1980 it is estimated that because of the trend to roll stock in existing grades, you will have to lay off half of the 100-man sheeting operation and shut down two cutters.

To: Marketing Manager
SUBJECT: FOIL LAMINATED PAPERS
Date: May, 1968

The market for this paper was estimated at $75,000,000 in 1965 and is growing at the rate of 20% per year. The price for the present paper is $600/ton and is sold by the manufacturer directly to users of high-speed printers. A paper that would perform satisfactorily could be priced at $700/ton.

We believe a realistic market share would be 10%, which we could quickly achieve and keep.

Presently, we do not have a product like this in the line. This means we would need to develop a new chain of distributors. Sales is not too anxious to do this at their present manning level, and they believe service calls would be frequent. Selling expense is estimated to be 20% of gross sales.

Freight must be prepaid, and would be $50 per ton on a product in this freight class.

To: Manufacturing Manager
SUBJECT: FOIL LAMINATED PAPER
Date: May, 1968

WISTU does not presently own any production-sized laminators. They are readily obtainable from commercial sources, however, at a cost of $2,000,000 each, installed. These have a capacity limit of 3,000 tons each, based on a 6-day week.

The equipment supplier has provided us with information on machine speeds, repairs, and maintenance costs. Knowing these, our labor costs, overhead, and raw material expenses, we have estimated a manufacturing cost of $300/ton on 60-inch wide rolls, and $400/ton on 40-inch rolls. The market is split about 50-50 on these roll widths. Although we have asked the supplier for a quotation on a 40-inch machine, he has responded to our inquiry by advising that production of a machine that narrow does not fit their present manufacturing capabilities.

FOIL LAMINATED PAPERS
R & E

Recently one of the scientific journals reported that the aluminum industry had made a large research grant to a mideastern university to study vacuum deposition of aluminum on film to provide a new high-speed printer communication medium. If successful, they believe the product would be sold for $500 a ton.

Trials on our pilot laminator indicate that with the added processing device on which we have patents, we can make a product that will command $700 per ton against existing competition.

A manufacturer of high-speed printout devices announces plans to begin a rather large research effort to develop a more efficient printer that could utilize ordinary paper.

FOIL LAMINATED PAPERS
Planning

It would be advantageous to install the foil laminating equipment in our East Coast mill, since it is closest to the customer, and freight is already $50/ton. However, the power requirements are such that a $500,000 addition to the mill's generating plant would also have to be installed for the new equipment.

If base paper being used for this project were to be made in the East Coast mill, several grades now being made there would have to be shifted to the Midwest mill, making them less competitive in Eastern markets.

        To: Marketing Manager
SUBJECT: DISPOSABLE PALLETS
      Date: May, 1968

In 1967, sales of all types of pallets were approximately $300 million. Rate of increase is projected at 5-10%

per year. The expendable type is 30% of the market and is growing twice as fast as nonexpendables. Nearly all expendables are hardwood, and they now cost approximately $13.00 each. Because of diminishing supply, hardwood prices have trended steadily up for the last few years.

Market research indicates a disposable pallet, which would hold 500 lb and sell for $10.00, would be well received. It is estimated that such a pallet would give us a 50% share of the present expendable pallet market plus a 10% share of the existing reusable pallet market.

Sales estimates costs for selling a product of this nature would be $1.00/unit. Freight charges would average 20 ¢ /unit.

To: Manufacturing Manager
SUBJECT: DISPOSABLE PALLETS
Date: May, 1968

Your pallet is made from fibers and adhesives. The pilot production unit you built for $100,000 confirmed estimates that a $7 million plant could turn out 2 million pallets a year, at a manufacturing cost of $5.00 per unit, including freight.

This plant would have to be built from scratch and the Vice President in charge of marketing has volunteered to turn the first spadeful of earth!

DISPOSABLE PALLETS
R & E

You have just learned of research underway by a competitor on adhesives to replace nails and make soft woods suitable for pallets. If successful, this would reduce wooden pallet prices, and probably cut your market share in half!

Our own research indicates our present design may be over-engineered, and an analysis, which has as its objective cutting the manufacturing cost of our pallet by 20%, is being made.

DISPOSABLE PALLETS
Planning

If we want the new pallet plant to be located near its raw material sources, it should be in the Southeastern U.S.A. If we want it near its customer, it should be in the Pacific Northwest.

One of our junior planners has suggested building the plant smaller than proposed but building two plants instead of one, thereby giving us some flexibility and insurance against fire and rail strikes.

PROJECT	A	B	C
	(Book Publishing)	(Foil)	(Pallet)
Gross Sales	$21,600,000	$12,250,000	$66,000,000
Less Freight	1,800,000	875,000	1,320,000
Net Sales	19,800,000	11,375,000	64,680,000
Less Mfg. Cost	14,220,000	6,125,000	33,000,000
Gross Mfg. Profit	5,580,000	5,250,000	31,680,000
Less Selling Expense	2,160,000	2,450,000	6,600,000
Gross Profit	3,420,000	2,800,000	25,080,000
Less Tax (50%)	1,710,000	1,400,000	12,540,000
Net Profit	$ 1,710,000	$ 1,400,000	$12,540,000
ROS	8.6%	12.3%	19.4%
ROI	19.0%	11.2%	44.7%

FINANCIAL

Because of the tight money markets presently existing, prime borrowers must pay 6% interest. All year WISTU stock has been selling for $25 per share. Recently the stock dipped to $22 and has not recovered, because of a slack period in the over-all economy.

WISTU'S financial analysts compute return on sales and return on investment in the following manner:

$$ROS = \frac{Net\ Profit}{Net\ Sales}$$

$$ROI = \frac{Net\ Profit}{Net\ Sales} \times \frac{Net\ Sales}{Investment}$$

Financial Department Policies require that all new product proposals be submitted in the following format:

Project	A	B	C
Gross Sales*			
Less Freight=			
Net Sales			
Less Mfg. Cost=			
Gross Mfg. Profit			
Less Sell. Exp.=			
Gross Profit			
Less Tax (50%)=			
Net Profit			
ROS=			
ROI=			

*To compute gross sales take size of total market times market share times selling price. (Normally, a project life would be estimated for several years, sales growth plugged in, and the sales and costs discounted for their time value). For this exercise assume a one-year life, and ignore the above refinements.

BIBLIOGRAPHY

Berg, Thomas L. and Schuchman, Abe (Editors) *Product Strategy and Management*, Holt, Rinehart and Winston, Inc., New York (1963)

Booz, Allen & Hamilton, Inc. *Management of New Products*, New York.

Corey, E. Raymond, *Developing Markets for New Materials*, Cambridge, Massachusetts, President and Fellows of Harvard College, (1956)

Hainer, Raymond M., Kingsbury, Sherman., Gleicher, David B., *Uncertainty in Research Management and New Product Development*, Reinhold Publishing Corporation, New York (1967)

Hilton, Peter, *Handbook of New Product Development*, Prentice-Hall, Inc., Englewood Cliffs, N.J. (1961)

Johnson, Arno H., Jones, Gilbert E., Lucas, Darrell B., *The American Market of the Future*, New York University, New York (1967)

Marting, Elizabeth (Editor), *New Products/New Profits*, American Management Association, Inc., New York. (1964)

Marvin, Philip, *Planning New Products*, (Second Edition), American Management Association, New York. Inc., (1964)

Pearl, D.R., *Creating Successful New Products*, an ASME publication, 66-WA/MGT-3, 1966.

Sullivan, Fred R., "Where are Tomorrow's Markets", *Duns Review*, November, 1967.

chapter 2

Marketing Research In The Specialty Paper Business

JACK M. RUDOLPH

INTRODUCTION AND OBJECTIVES

Because the primary readers of these volumes are those technically oriented people who are producing or considering the production of Specialty Papers, a considerable number may have little familiarity with the real importance of Marketing Research. They probably have had little exposure to what it is, what it can do and what it cannot do; they probably have given little thought to how they might put it to use.

We cannot attempt with a few pages to produce experts in marketing research techniques. That is a job for colleges, universities, and industry—and for the numerous special seminars on this subject sponsored regularly by various management, trade, and professional associations. On the other hand, for those who might want to dig deeper after this brief introduction to marketing research we have a bibliography at the end of the chapter that includes some selected textbooks and other reference literature providing detailed information on procedures and techniques.

Rather, the objectives of this chapter are:
1.   To encourage the reader's demand for more and better information about the requirements and desires of his customers. (Note that it is the function of good marketing research to provide just such information!) The approach to this objective will be a related review of the economic conditions and the competitive developments within the total business environment that have made the application of such information a critical factor for successful business management.

2. To encourage the reader's ready willingness to make use of marketing research. We hope to give him a "comfortable" familiarity with marketing research, and an ability to communicate his needs for and use of it. The approach here will be a general discussion of the characteristics and capabilities of marketing research, an introduction to the language, procedures and techniques used, and a quick exposure to the many kinds and applications of it.

3. To give the reader some helpful guideposts for managing or conducting marketing research to meet his specific information needs. This will be done by separate specific commentaries interspersed throughout the general discussion. These will attempt to present key elements of the marketing research process in enough detail to provide the reader with a working knowledge of those elements. We hope these "guideposts" on various basic facets will equip the reader to:

   a. Identify, break down, and define that marketing information needed for specific decisions—*the first requirement for insuring effective marketing research.*

   b. Evaluate whether the research can be done by internal capabilities or whether it requires expert external services.

   c. Distinguish between competent and incompetent researchers.

   d. Distinguish between "good" research and "bad" research.

   e. Conduct, personally, or direct other internal personnel in, "good" marketing research of an uncomplicated nature that will probably satisfy a large portion of his information needs.

The specialty paper business is characterized by a higher than average rate of product obsolescence. Rapid technological developments, particularly in the fields of resins and plastics, are increasing this rate. They are at the same time, however, increasing the potential for new products and new market applications. The marketing information needs in specialty papers are, therefore, heavily oriented to those business decisions that are connected with new product development and introduction. The discussions in this chapter take that into account, highlighting those types, sources, and applications of marketing information, and emphasizing those kinds of research and procedures most frequently encountered in development of new products.

On the other hand, the markets for Specialty Papers are perhaps more widely diversified than those for most other industries. Somewhere in the maze of markets and products, decisions concerning every marketing function, every kind of distribution, and every kind of customer are involved. Thus, a general coverage of the whole field of marketing research is most appropriate.

THE URGENT CASE FOR MARKETING RESEARCH

The use of deliberate and effective marketing research has become imperative for the success or survival of any business enterprise.

From all indications there will continue to be an accelerating rate of change in the competitive environment for every business and in the product-service-sales requirements for every market. A few of the factors forcing this change are:

a. A continuing development of increasingly sophisticated management techniques and tools—e.g. computerized business information systems, mathematical models for business simulation on a computer, Bayesian Decision Theory, PERT/CPM (Program Evaluation Review Technique/Critical Path Method), Input-Output Analysis, and Ratio Analysis.

b. A growing number of highly qualified and truly professional management personnel, competent in the use of these new techniques—managers that are using objective, market-oriented, product- and business-planning programs.

c. An expanding base of technology that is providing an increasing volume and variety of new products, new processes, and new business techniques.

d. The development of increasingly effective communications facilities and skills that help introduce and exploit new products or marketing innovations more rapidly.

e. Increasing business diversification both by internal development and by acquisition.

On the other hand, we have increasing limitations and demands being made by labor and government. These increase costs of product development and operating capital. The financial consequences of being wrong in the marketplace with any business decision have become critical, but this is particularly so with respect to the heavy development and introductory expenses necessary to bring out new products.

The National Industrial Conference Board has made a number of studies concerning the experience of companies with the development and market introduction of new products.[1,2,3,16,21,23,24,27,28] One of the studies shows that an average of 3 out of 10 new products failed[24] even after being fully developed and introduced to the market. Most of the reasons given for failures of new products stem from inadequate knowledge of the requirements of the market[16,28] including:

1. A lack of knowledge of, or an imperfect understanding of the product features or functions required by the prospective users,
2. Inaccurate estimates of the present and future volume requirement for the product,
3. Improper evaluation of the rate at which prospective customers would convert to, adopt or install the new product,
4. Failure to determine what distribution channels should be used or a lack of knowledge of the costs of distribution, and
5. Insufficient knowledge regarding competitive products, practices, performances and developments.

Businesses can no longer afford the cut-and-try approach to the market that was adequate as recently as 15 to 20 years ago. To remain financially solvent today, an enterprise must predetermine its chances of a profitable fit to the requirements of the market before it commits any significant amount of manpower or capital resource to a new business development—or even to a continuation of its existing business.

GENERAL DEFINITIONS AND APPLICATIONS OF MARKETING RESEARCH

Finding out the requirements of the market to evaluate its chances of a business fit—the determination of what is the market, *what* it wants, *when* it needs it, *where* it needs it, *how much* it needs, *why* it is wanted, *how intensely* it is wanted—this is marketing research.

In the Definitions Manual of the American Marketing Association, Marketing Research is defined as "the systematic *gathering, recording* and *analyzing* of data about problems relating to the marketing of goods and services."[-66,68]

The emphasis of modern professional business management on the concepts of "Marketing Orientation" and "Management by Objectives" point up the importance of marketing research in successful business management.

The concept of "Management by Objectives" emphasizes a systematic process of deciding on objectives to be accomplished; then progressing through the planning and implementation of the programs, organization, motivation and controls to reach those objectives. It is applicable to every function and activity of a business—at every level of management responsibility. *It starts with—and is centered on—a systematic collection and analysis of relevant information as the basis for sound decisions at every step of the process.* A good graphic con-

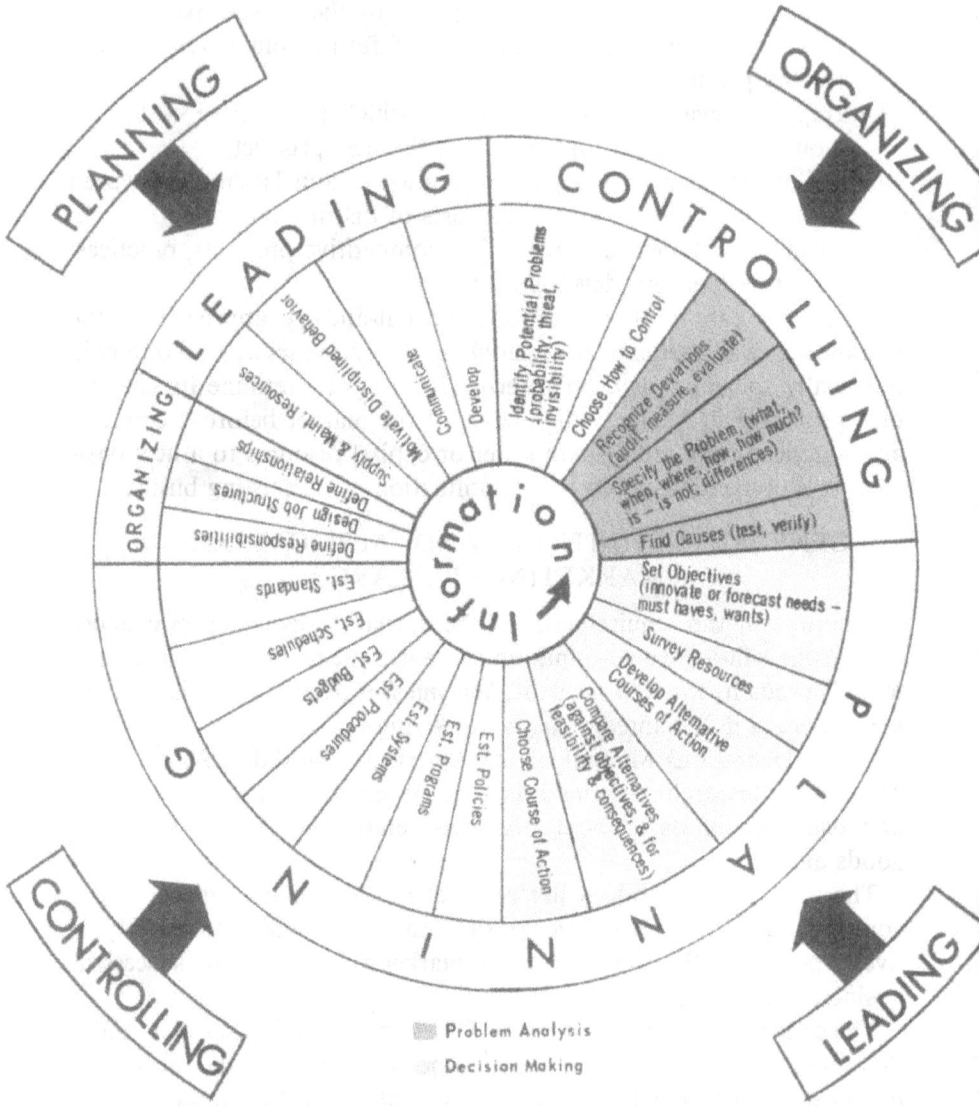

Problem Analysis

Decision Making

THE MANAGEMENT WHEEL

Fig. 2-1. *Courtesy the Management Research & Development Dept., Kimberly-Clark Corp.*

ception of the process is represented by the "Management Wheel" reproduced in Figure 2-1. Note that the hub of the wheel—the basic core around which the whole managing process revolves—is "IN-FORMATION"—objective, relevant, factual information.

The concept of "Marketing Orientation" has been defined as "The total management of a business that makes sure the *needs and desires of its customers are identified, understood and considered* in *all* its business decisions."

The implications for marketing research presented by these two management concepts are obvious. The systematic collection and analysis provided by good marketing research does produce information that is relevant and necessary for good business management decisions; and the systematic recording and reporting can help assure consistent communication and understanding of this information at all levels of the organization that should be involved.

PROFESSIONAL MARKETING RESEARCH

Marketing Research is a profession. There are four-year college curriculums in marketing with a major specialization in Marketing Research. There are associations of Marketing Consultants and Researchers with published codes of professional practices and business ethics, which the members are expected to follow. There are practicing marketing researchers qualified by specialized training or experience to do an expert professional research job. Unfortunately there are also some incompetent promoters.

When the need for marketing information is critical, or the research required to develop it is complicated, it is generally wise to have an experienced expert on the job. How can the expert be identified? One looks for evidence of the basic characteristics of good research (discussed later in this chapter) in previous work he has done. One listens for it in preliminary discussions of the problem with him. One asks what others have employed him for a similar job and talks to them about his performance. Depending on the scope of the job and the importance of the information, it may be desirable to prescreen two or three possible candidates, ask them for written proposals on how they would attack the job, and compare the proposals. The writer recommends (1) the business research report, "Using Marketing Consultants and Research Agencies"[14] published by The National Industrial Conference Board for guides in locating, selecting and working with outside professional marketing researchers and (2) "Selecting

Marketing Research Services," Management Aids Annual No. 9, chapter 10, published by the Small Business Administration.[63]

There is "good" marketing research, producing specific, relevant, honest information and there is "bad" marketing research, producing general, unrelated information or misleading fantasies. A decision maker needs to be able to distinguish between them, to insist on getting and using the former and to expose and reject the latter. Flying without instruments is dangerous—trusting in faulty instruments is suicide!

PRACTICAL EMPLOYMENT OF MARKETING RESEARCH

There is no magic or mysticism in good marketing research. Neither is marketing research a precise science that always requires a highly trained technician. It ought to follow the general procedural steps that are widely accepted and recommended for any investigation of facts, often referred to as "the scientific method" (see later section outlining this method)—and it can benefit greatly from the use of precise mathematical techniques for testing sample reliability, for organizing and analyzing the data collected, and for evaluating the significance of results. However, the real basic requirements for all good marketing research are complete objectivity, organized logical thinking, good common sense, innovative imagination, and conscientious application of actual detailed "pick-and-shovel" digging, sorting, comparing, and reporting. The point is that the reader, or any responsible person in his company, that can and will apply these five basic characteristics to the information development task, can produce sound effective marketing information adequate for the majority of business decisions encountered in the development and marketing of specialty papers. No doubt, some very competent professional marketing researchers might question such a statement, but over years of marketing research and marketing planning activities on paper and forest products for industrial and commercial uses, many intelligent business people, none of whom had any academic training or professional experience in Marketing Research have produced effective results. Most of the projects involved were not identified or thought of as marketing research; but they were nonetheless a systematic objective development of specific information about customers or potential customers that was needed to make specific business decisions. Quite often there were significant secondary benefits because the people involved became more intimately acquainted with the market as a result of their specific research activity.

NONOBJECTIVITY IN RESEARCH

Objectivity is perhaps the most important characteristic of good marketing research; a lack of it is the most frequent cause of bad or misleading research. A lack of objectivity may enter the research project at almost any point in the process from the initial request for research to the final reporting of results—and sometimes even in a misapplication of good results by the ultimate decision maker. The lack of objectivity may be completely unintentional and stem from the researcher's naivete and inexperience, or it may stem from his unconscious bias because of an overexposure to the market and/or the problem. It may be a conscious compromise of objectivity because of pressures of time or budget, or just plain laziness; it might even be a deliberate biasing of results to protect a preconceived opinion or position. There are two broad classes of nonobjectivity in marketing research— "irrelevant" research, and "biased" or nonrepresentative research.

In the 1966 Annual Marketing Conference of the National Industrial Conference Board, Donald Davis, President of The Stanley Works, who was moderator of a panel discussion on "The Contributions of Market Research to Better Planning," summed up one of the presentations with this very pertinent statement, "Marketing Research is effective *only* when it is undertaken to provide the *specific market information needed for a specific decision;* when it includes thoughtful analysis and recommendations *relative to the decision;* and when it is followed up to insure adequate understanding, interpretation and consideration in the making of the decision." He also went on to observe that the systematic formalized business planning procedures being used in most of the well-managed companies today had a beneficial feedback on the effectiveness of the marketing research functions in those companies, because such procedures force an identification and definition of *specific* information gaps early in the planning process.

Irrelevant Research

Much of the responsibility for irrelevant or misdirected marketing research must be assumed by the decision maker that originates the request for the research if he fails to identify the specific decisions he must make and the specific marketing information he needs. On the other hand, one of the characteristics of an objective approach to investigating any kind of a problem—and certainly one of the characteristics that should be expected from a professional marketing re-

```
MARKETING RESEARCH DEPARTMENT                    cc-
   XYZ CORPORATION                                (President)
                                                  (V.P. Marketing)
                                                  (V.P. Requesting Dept.)
REQUEST FOR MARKETING RESEARCH SERVICE            (Requester)
                                                  (Mgr., Market Research)

SUBJECT AREA/PROJECT TITLE:                       _____
                                                  _____
                                                  _____
‡
_____

OUTLINE OF PROBLEM (Including Pertinent Situation Background and Environmental
    Factors):




_____

MARKET INFORMATION NEEDED:



_____

DECISION APPLICATION INTENDED:



_____

PROPOSED PROCEDURE (Attach additional pages for detail.):
    Source and Type of Data:
    Method of Data Collection:
    Data Analysis Techniques:
    Sample Design:

    Resources Needed:
      Personnel
      EDP Facilities
      Materials/Forms



_____

ESTIMATED COSTS:   INTERNAL $_____     SCHEDULE:_____

                   EXTERNAL $_____     ASSIGNED TO:_____

REQUESTED BY:_____     APPROVED BY:_____
DATE:                                             DATE:
```

Fig. 2-2. Typical format for a marketing research plan.

searcher—is an insistence on a clear, mutually understood, definition of the problem, the information needed, and the way in which the information will be applied (the decision required), before starting to organize the final research plan. Sometimes this effort on identifying specific decision requirements will refocus the decision maker's understanding of the problem, and completely change his initial concepts of the information needed. It might even eliminate the need for any research.

Therefore, an identification of the specific information needed for specific decisions is the first rquirement for insuring objective, relevant marketing research whether the research is undertaken internally or sought externally. The precise specification of information needed and how it will be used are checkpoints for evaluating the competence of an outside professional. Clear, concise statements concerning the problem, the information required and the purpose for which it will be used should be the first part of a formal written outline of the research plan included in any proposal from an outside agency. In fact, such a formal written plan should be prepared whether the research is undertaken internally or externally. Figure 2-2 is an example of a typical format for a Research Plan outline.

Even when the decision and the information requirements have been clearly identified and specified, one may still need to audit and/or caution the researcher (particularly inexperienced internal staff personnel) against getting carried away with his assignment. Sometimes ambitious researchers, or researchers anxious to make a good impression, collect, analyze, and report a great deal of information in addition to that specifically required. Usually the additional information is unrelated to the problem or requirements of the decision. It can clutter up a quick identification and understanding of the relevant information. It can confuse the analysis and conclusions relative to the decision. It is probably a waste of time and money.

Nonrepresentative Research

The second broad class of nonobjectivity in marketing research is the production of results that are not representative of the real situation. This problem should be considered from a pratical recognition that little, if any, information about the marketplace is absolutely accurate and completely representative of the real situation. The most carefully planned and conducted collection of market data always contains some variation from the true situation. This is not a condemnation of marketing research, or a caution for the decision maker against using it. Rather, it points up the fact that the decision maker needs to (1) be aware of the primary causes for nonrepresentative research information, (2) watch for and minimize these causes to the extent possible in research he undertakes internally or employs externally, and (3) look for evidence of them and judge their possible effect on the reliability of *any* information he uses for guiding his decisions.

David Levine, Director of Market Intelligence for U.S. Plywood, summed up this situation quite well in his presentation at NICB's Annual Marketing Conference in 1966. He pointed out that decision makers must not expect market information to be precise, invariable fact, or to use it as such. Nevertheless, they must obtain and use the best possible market information for sound business decisions. He observed that Marketing Research is not a precise science, but that it can produce sound *relative* information about the market or about its reaction to a vendor's marketing efforts. Market researchers should always use the scientific method of investigation; whenever applicable, they should use the precise mathematical techniques and scientific theory that are available for sample design, questionnaire design, testing of data for consistency and analysis of data. Levine emphasized that decision makers should insist on the information they use being developed in a systematic, objective way.

The basic causes of nonrepresentative research results come from:
1. Nonrepresentative samples
2. Inaccuracy of individual data
    a. Wrong answers from individual respondents that are caused by poorly designed questions
    b. Errors in recording of raw data by researchers
3. Improper analysis of data
    a. Naive or superficial analysis
    b. Deliberate bias.

Sampling theory and the statistical analysis of sample data are too complex to attempt the development of any workable knowledge on the subject in this chapter. Some references have been included in the bibliography which should be helpful for that purpose.[95-103, inc.]

Nonrepresentative Samples

The considerations for selecting a research sample design, or for judging the effect of sampling error on the reliability of marketing information obtained from any source vary for each specific requirement encountered. If the business decision to be made relates to a specialty paper product that is custom-designed for a single industrial customer and "market information" about that customer is needed, there is no sampling problem. But suppose that a single customer uses the specialty paper product as a component in a consumer item that has wide national distribution, and information, which he doesn't know or doesn't disclose, about the market acceptance and future of the customer's product is needed. In this case, a proper sample design could be very important for getting objective reliable information.

For many information needs, a statistically representative sample of the market is not critical. This is generally so in the product concept and the product performance testing associated with new product development. On the other hand, it is always questionable, and sometimes dangerous, to project design preference, final product acceptance, or volume expectations for the total market on the basis of research results from a nonrepresentative sample. The following are examples of questions that ought to be asked about the sample when considering the objectivity of research performed, or the reliability of information obtained:

— Is it critical whether this information truly represents the total market?
— Was the sample used statistically representative of the total market?
— If not, is this a good profile of what it does represent and how that relates to the total market?
— Was this considered in analyzing and reporting the information?
— Is there any evidence of deliberate bias in selection of the sample used?
— Was the sample large enough to give significance to the information that was generated?

Inaccuracy of Data

A certain amount of nonrepresentative research that results from errors in the recording or processing of raw data by careless, nonobjective, or dishonest interviewers and clerks always exists. The reader should know that this can and does happen occasionally, particularly where independent part-time interviewers are employed and where nonprofessional clerical personnel are used in the processing and recording of data. This situation is difficult for a research client to audit or to detect in the final report. Suffice it to say that most major operations for collecting data—e.g. the Bureau of the Census, professional marketing research agencies or large company research operations—have routine systems, which are sometimes quite elaborate, for checking a certain random sample of the work of field interviewers and of clerical personnel that process data. The reader may want to explore the provisions for this in checking on the qualifications of external research services.

The most frequent cause of inaccurate raw data probably stems from poor design of questionnaires. The reader does have some chance to audit this factor in research obtained from external sources; he can certainly control or influence it in internal research. Interrogatory technique and questionnaire design is a science in itself. Some principles and theories that apply have had a long history of development in the field of law. Most Marketing or Marketing Research textbooks and reference books cover the subject at some length. This chapter lists ten basic rules for questionnaire design as a guide for the reader:

1. Questions should ask *only* for data that can be *clearly remembered* by respondents, or that is *readily accessible in records*. A caution to keep in mind here is that people tend to *guess at* data rather than admit they don't remember—and this can distort results.

2. Questions should ask for reports of specific events rather than generalizations. This includes, so far as possible, an avoidance of requests for off-hand estimates of percentage splits and average monthly volumes.

3. Questions should be simple, clear, and direct. The one possible meaning should be obvious to persons of all intelligence levels. Words subject to different interpretation should be avoided (e.g. "What kind of label stock do you use?"—What is meant by *kind*? Rolls or sheets? Coated or uncoated? Gummed or plain?) When possible, avoid or explain exactly what is meant by any technical terms or descriptive "language" peculiar to a specific industry, trade, or profession.

4. Questions should not be leading. This topic has many facets, but it includes (a) questions worded so that respondents usually avoid one of the two possible answers because it "puts him in an unfavorable light", (b) questions worded so that respondents usually answer one of the two possible answers because it appears to be the one desired and "he wants to be a good guy", (c) multiple choice check-off answers which tend to "stimulate imagination" (d) even more subtle questions to which previous repetitive mentions or question sequence "suggest" an answer that would be "favorable" to the interviewer.

5. Questions that raise a personal bias or prejudice should be omitted (unless this is the direct objective of the survey!)
6. Questions should be limited as much as possible to facts and attitudes. Questions asking for interpretive answers, particularly for the reasons for respondent's actions or attitudes, are difficult for people to answer *correctly*. (This requires a specialized type of investigation or research.)
7. Questions should be as easy to answer as possible—physically as well as mentally and emotionally.
8. Questions containing more than one element should be eliminated.
9. All questions should provide for *conditional* answers rather than "forcing" respondent to fit his experience in a yes/no situation. Questions should provide for a "maybe" or "don't know" response—*these should be accounted for in the analysis.*
10. The questions should be arranged in the *proper psychological sequence*—not necessarily the order of logical development of ideas or the best order for tabulating and analyzing! This usually means that the first questions should be interesting and easy for the respondent to answer, generating a momentum of willing response, which carries him through succeeding questions that become progressively more difficult for him to answer.

Unrecognized "violations" of many of these "rules" often show up because of subtle influences in a particular survey situation—even in questionnaires designed by experts with a great deal of experience. Pretesting of questionnaires with a pilot group of respondents is always advisable. In this phase of questionnaire development a careful observation should be maintained for facial and voice tone reactions, and consistent patterns of responses to individual questions. Be suspicious of questions that turn up the same answer consistently—or of questions on which most respondents indicate some difficulty or reluctance to answer. The rules and practices that have been reviewed here apply to questionnaires used in industrial market research as well as consumer market research.

Improper Analysis of Data

With regard to nonrepresentative information resulting from improper analysis of data, it is probable that most of the marketing information needs that are encountered by managers in the specialty paper business seldom require complicated research plans involving the collection of masses of raw data from rigorously controlled sampling structures. In other words, research that requires sophisticated analytical techniques or complicated statistical manipulation of the data for accurate analysis is seldom necessary. This chapter does not attempt to provide a working knowledge of statistical data analysis. References that present detailed explanations and applications for various mathematical techniques commonly used in analyzing research data and in testing the significance of results appear in the bibliography.[77,98 −103 inc.] Another chapter in this volume touches on some of the newer, more sophisticated computer simulation and Operations Research Techniques. These are being used more and more in the analytical activity connected with marketing research.

For the objectives of this chapter, it is sufficient that the reader recognizes that "biased" research information can result from the analytical phase of the research, either naively or deliberately. Although evaluation of research results for adequate analytical competence without a working knowledge of the statistical analysis techniques may be difficult, two checkpoints that can usually be detected quite readily are:

1. Were all the samples and all the data collected (including "don't know" or no-answer responses), "accounted" for, and recognized in the analysis and reporting of results?

2. Is there evidence that concerted attempts were made to develop segmented breakdowns and relationships within the data rather than masking everything into broad averages—particularly where there is considerable variation between individual data?

THE SCIENTIFIC METHOD OF INVESTIGATION

In this chapter frequent references are made to the use of "The Scientific Method" for the development of needed marketing information. What is meant by this term? What does it involve?

Over the years, scholars of human accomplishment have discovered that people who developed a reputation for solving problems or for making new discoveries appeared to follow similar systematic patterns of thought and effort. These patterns were studied, specific phases or steps were isolated and defined, and then the steps were arranged in the sequence in which they normally occurred. This identified system of thought and effort became known as "The Scientific Method" because it was originally associated with the early giants of scientific discovery. Today it could just as well be called "The Engineering Method" or "The Investigative Process" or "Systematic Problem Analysis" or "The Theory and Logic of Research."

Different writers have listed a larger or smaller number of steps in the sequence depending on whether they combined or separated certain elements, but the general sequence of activities has stood the test of time.   It is applicable to the Marketing Research task; and the researcher who disciplines his thoughts and efforts in this systematic manner is consistently more effective than one who does not.   The following is "one" list of the sequence of steps in "The Scientific Method."   Alternative titles for the various steps, as used by different writers or in different contexts, have been included for the possible insight and guidance they might provide.

THE SCIENTIFIC METHOD

1. *Identify and Define the Problem.*—Situation analysis—Survey of environment—Background analysis—Pilot study—Informal investigation

2. *Define the Purpose of the Research.*—Establish the hypotheses—Specify the decision to be made—(Also combined with Pilot study—Informal investigation)

3. *Specify the Information Needed.*—Define the objectives—Specify the tests of the hypotheses—(Also combined with Pilot study—Informal investigation)

4. *Survey for Type and Sources of Relevant Data.*—Survey of literature—Library and records search—(Also combined with Pilot study—Informal investigation)

5. *Consider Research Alternatives and Select Approach.*

6. *Prepare Detailed Plan of Formal Investigation.*—Establish the program of action.   Plan and organize the research.
   a. Select the types and sources of data to be collected.
   b. Select the methods of collecting data.
   c. Select the methods and techniques for analyzing data.
   d. Establish sample design.
   e. Prepare (and test) forms (questionnaires) for collecting and recording data.
   f. Determine organization and training required.
   g. Estimate time and cost.

7. *Collect and Record Data.*—Conduct the formal investigation—Conduct the experiment and record the observations (Note—This step would include assembling and training any additional personnel required.

8. *Analyze Data.*—Organize and process the data.   Examine the data for interrelationships.

9. *Interpretation of Results.*—Develop conclusions and recommendations.

10. *Preparation of Report.*—Communicate the results.

Practically all recent writings in the field of marketing management and marketing research add a final step to this system as it applies to marketing research, i.e. "Follow Up to Insure Proper Application."

A considerable amount of discussion about some of the elements of this research method has already been presented in the section on nonobjectivity—and some additional detail on "methods of collecting data" and on "sources of data" is given in subsequent sections of this chapter. Beyond that, the reader is directed to detailed discussions in the Marketing Research textbooks and reference books listed in the bibliography [57,84-94 inc.], if he needs further insight in putting this "method" to work.

The amount of effort required under individual steps of the above outline varies for each research task encountered. As previously pointed out, the decision maker should have already accomplished much of the first three steps on this list when he asks for research. Nevertheless, the researcher still needs to become thoroughly familiar with the problem background, and to have a complete and mutually agreeable understanding of the specific decision requirement and the information needed to mount an efficient and objective research project. Frequently, also, the step 4 "Survey for Type and Sources of Relevant Data" may turn up available published information that satisfies the objectives of the research project with little or no further work except rearranging its form or relating it to the specific problem or decision. Still, it is good practice for any researcher to think every research project through each step of the "method."

SOURCES OF MARKETING INFORMATION

To list all varied sources from which marketing data or marketing information are obtained is impossible. Some general discussion about kinds of data sources is presented; then a few selected secondary sources that have been found most frequently helpful in searching for specialty paper marketing information needs is listed. Some fairly comprehensive sources of information, guidebooks, catalogs, and directories,[32,33,49,67,70,74,134] are listed in the bibliography at the end of this chapter. Many of the other references also contain extensive bibliographies of sources of marketing information.

Sources of marketing information are broadly classed as *primary*, or original, and as *secondary*. Primary sources are the individual people or companies directly involved in a specific marketing situation. They are in turn broadly classed into *suppliers* or producers, *dealers*, and *customers*. Secondary sources derive from the activity of institutions or individuals that are generally external to the specific marketing situation; the information provided is often only indirectly related. The possibilities are so broad and varied that all-inclusive subclassification is difficult. Some of the more common general categories include *government* statistics[49,135] (federal, state or local), *association* statistical and information programs[53,113,134], the editorial or promotional research of *communications media*[66,133] (newspapers, magazines, radio and TV broadcasting companies) and *research organizations*[104,105,107,132] (both nonprofit and commercial.)

Primary Sources

A company's own files and internal business records are rich primary "supplier" sources of specific marketing information that are quite often overlooked. They are frequently the best possible sources—and sometimes the only sources—for an identified list of known dealers or users needed for a specific research project. They should always be examined for ways in which they might be more effectively organized for exploitation as original marketing information sources. The company's own sales and marketing personnel, who are in daily contact with the market, are excellent primary sources that can be tapped for specific information. The sales personnel or general management personnel of other noncompetitive suppliers to the specific market being studied are often fruitful primary information sources. Even directly competitive suppliers are legal and ethical sources for some kinds of information.

Several distinct categories of "dealer" operations might be involved in any specific marketing situation including *Retailers*, *Wholesalers*, *Jobbers* or *Agents* of various kinds—any middleman functioning directly in the transfer of the specific product or service from the producer to the end-use customer. In addition to the different kinds of establishments that might be involved here, is the possibility—and for certain research objectives often the desirability—of further distinguishing between *dealer management* and *dealer sales personnel* in this class of primary sources. Of course any internal business records of the dealers involved fall into this same class. "Customer" sources can and often should be distinguished between *buyers, buying influentials,* and *users*.

In general, primary sources will be required to satisfy a major part of the marketing information needs in the areas of:

1. Identity and Classification of the nature and characteristics of potential customers.
2. Product/Price Concepts or Acceptance—Fit to potential customer's design/performance/value requirements.
3. Potential Usage Patterns—Specific data on product volume per "capita" per time period for different classes of customers or potential customers.
4. Specific Product or Company "Image" Measurements at customer and/or dealer level (although there is some work of this nature being done on a routine basis and available as "secondary" source information—at a price—from communications media and from independent commercial research organizations.).
5. Evaluation of Alternative Marketing Strategies or Programs.
6. Analysis of Distribution Costs and Profit Contribution.

Secondary Sources

Secondary sources are generally thought of as publicly available *published reports and information,* usually of a statistical nature and often issued on a routine periodic basis. These are compiled by trade associations, communications media, government agencies, and universities. However, the staff personnel of the organizations that publish such material are also excellent "secondary" sources that should always be considered.

Trade Associations

A researcher should always contact the administrative staffs of all the trade associations representing suppliers, dealers, or users involved in the marketing environment in which he is seeking information— and he should do this early in the preliminary situation survey of each project he undertakes. Generally, the detailed current reports and statistics compiled by trade associations are available only to the participating members of the association. However, summary information, historical information and counseling guidance on specific information needs are normally made available to nonmembers or external interests.

Our paper industry is particularly blessed with a wealth of secondary statistical information about installed production capacity and planned new capacity; operating rates (percentage of capacity); volume of order receipts, production, shipments and order backlog; shipments by grade and by end use; geographic destination of shipments; economic reviews and forecasts; and many other useful types of business information compiled by the various associations of manufacturers of pulp, paper, paperboard, allied products. Numerous reports are compiled and issued on a regular periodic basis, weekly, monthly, quarterly, or annually depending on the subject and the frequency needed by the members participating in the particular programs. The various manufacturers that compriae the American Paper Institute probably generate more association statistics than are available for any other industry except possibly steel.

Communications Media

Even though none of the associations contacted proves to be a direct source of relevant secondary information, the administrative staff personnel are usually well informed in their particular field of interest and can frequently guide the researcher very quickly to existing sources for the needed information or they can provide good counsel on ways to develop it. The same is true of the staff personnel of the communications media, particularly the editorial and research staffs of trade magazines specializing in fields related to the market situation on which the researcher is seeking information. For this reason, two of the more important sources on our selected list are *"Encyclopedia of Associations, Volume I"*—Published by Gale Research Company, Detroit, Michigan; and *"Business Publication Rate and Data"*—published by Standard Rate and Data Service, Inc., Skokie, Illinois.

One word of caution is in order concerning secondary source information generated by communications media. Although these organizations usually have capable, well-informed research personnel, and their information for editorial purposes is usually objective and reliable, all too often the market information that they circulate for their own promotional purposes tends to have considerable bias. The user of secondary information that is obtained from trade magazines or from any other communications media source should be particularly diligent in examining the objectivity of the research behind the information.

Govermment Agencies

Government agencies are indispensable secondary sources of statistical data for marketing research. Government agencies at all levels—federal, state and local—have been expanding their fact-gathering and marketing-research activities, with considerable encouragement from private interests in many sectors. Individual private firms, or even large well-financed associations, cannot hope to duplicate the fact-gathering facilities or the legal authority for obtaining answers that are available to government agencies, particularly those at the federal level. In general, the sample coverage and the checking and control of raw data collection in government statistical programs is excellent and the information compiled meets high standards of objectivity and reliability. Many different government agencies have fact-gathering and statistical programs in their own areas of interest that often prove valuable for particular private informational needs, but the programs of the U.S. Department of Commerce, Bureau of the Census are by far the most frequently useful.

A researcher should gain some familiarity with the basic statistical programs and with the general kinds of statistics produced by all the various government agencies. A federal government organization manual published annually by the General Services Administration, and a monthly catalog supplied by the Government Printing Office list all the material published by all federal government agencies. Both can be generally helpful for this purpose. Similar published outlines of organizational structure and programs are usually available for state governments, and for many local governments.

The researcher, however, should become particularly familiar with the basic programs of the U.S. Bureau of the Census, and with its published statistical material. The Bureau of Census publishes a quarterly catalog listing all material published by the Bureau. Although we have included some of the individual statistical programs of the Bureau of the Census in our selected list of secondary sources, we have also included the "Bureau of the Census Catalog", a key item on the list.

Some years ago to coordinate the statistical programs of the various agencies of the federal government, and to put their data collection and compilation on a comparable base so that cross comparisons and integration of data could be accomplished, an Office of Statistical Standards was set up in the Bureau of the Budget, Executive Office of the President. This office organized a Technical Committee on Standard Industrial Classification which undertook the task of developing a standard classification and definition system to encompass every kind of business and institutional establishment that existed in the U.S. The result of their effort was what we know today as the SIC system— Standard Industrial Classification System.

Its use for identifying, classifying and compiling data concerning business and industry was adopted by all federal government agencies more than 20 years ago. Most private research organizations and compilers of business directories also use that system. Any researcher or user of statistical information needs to be familiar with the SIC coding and definitions system to understand, interpret, and properly use the government statistics or most other available secondary information.[136]. Thus, another key item in our selected list of secondary sources is the "Standard Industrial Classification Manual—1967 Edition," published by the Bureau of the Budget, Executive Office of the President. Along with this goes a standard coding system, with definitions, developed by the Bureau of the Census for classifying and identifying all industrial *products*. This system is based on and coordinated with the SIC system for classifying industrial establishments. The latest revision of this was prepared for use in the 1967 Census of Manufactures and is called the "Numerical List of Manufactured Products—1967 Census of Manufactures.

Universities

The marketing departments of colleges and universities offering degrees in Business Administration or Economics often have extensive and continuing programs for developing marketing information. Sometimes they have specialized Bureaus of Business Research that undertake major studies of specific markets. Contacts with the heads and staff personnel of these departments are usually fruitful and may provide the specific information needed or valuable assistance in locating sources for it. Researchers should always take advantage of the opportunities for research assistance that may be available in local colleges and universities. Sometimes university staff personnel undertake private research projects for individual firms.

Some use of secondary source information is appropriate in practically every research project undertaken—even if it does no more than identify specific individual primary sources. Frequently the researcher finds information from secondary sources that completely satisfies his project. In general, information from secondary sources might be expected to satisfy a major part of the marketing information needs in the areas of:
1. Total Market Potential.
   a. Total number of potential customers (if the nature and characteristics of users have already been identified.)
   b. Total volume potential (if probable share relationships within

a standard product class are known—or if probable usage
patterns by classes of customers have been determined.)
2. Growth Patterns and Forecasts of Total Market Potential.
3. Geographic Distribution of Market Potential.
4. Lists of Potential Customers identified by name and address
   (from Dun & Bradstreet Customer Data Files or other commercial
   "list" sources, state industrial diretories, classified sections of
   telephone directories and other similar types of directory sources)
5. General Nature and Characteristics of Distribution Channels,
   including lists of potential "dealers" identified by name and address.
6. General Nature and Characteristics of Competitive Industry
   Structure.
   a. Number, size classification, and geographic distribution of
      plants producing similar class of products.
   b. Broad financial performance ratios within the industry.
   c. Lists of competitive suppliers by name and address (from source
      of supply directories and similar listings.)
7. Identification of specific Primary and Secondary Sources of Information.

Sometimes information that satisfies needs the of decision makers
in the following areas is available from secondary sources; otherwise,
primary sources must be used:
1. Product Purchasing Patterns—Sales Contact and Service Needs
   and Distribution Channel Requirements.
2. Purchasing Motivations.
3. Dealer Sales and Service Requirements—and Motivations.

The following is a rather arbitrary selection of secondary information sources that can serve as a basic starting point for researching
most of the marketing information needs that are encountered in a specialty papers business:

Standard Industrial Classification Manual

Published by the Executive Office of the President, Bureau of the
Budget, Office of Statistical Standards. Last edition available at time
of this writing was 1967. Previous edition had been 1957, with a
1963 supplement of additions and changes, $4.50.

The following six sources are compiled and published by the U.S.
Dept. of Commerce, Bureau of the Census:

Numerical List of Manufactured Products

Usually revised and published in conjunction with the quinquennial

Census of Manufactures.   Last revision at the time of this writing was made in 1967 for use with the 1967 census.

Census of Manufactures

A collection and compilation of detailed statistics on number and location of plants, employment, volume, and value of production by individual kind of products, payroll costs, material costs and value added by manufacture, capital expenditures, and various other miscellaneous information covering each SIC classification of the manufacturing industry.   Individual reports are issued on individual subjects, kinds of industries and products, and geographical areas and finally assembled into a set of bound volumes.   The census is made at nominal five-year intervals; the last one for which final reports are available at the time of this writing is for 1963.   The next previous censuses in the series were completed for 1958, 1954, and 1947; one is currently in process for 1967.

Census of Business

A collection and compilation of detailed statistics on numbers of establishments, employment, volume of business for each SIC classification covering Retail Trades, Wholesale Trades and Selected Service Trades.   Again, individual reports are issued on individual subjects, kinds of business, and geographical areas—and they are finally assembled into a set of bound volumes.   This census is also made at nominal five-year intervals; the last one for which reports are available at the time of this writing is for 1963.   Immediately previous censuses had been completed for 1958, 1954 and 1948; one is in process for 1967.

Census of Population

A collection and compilation of detailed statistics covering a large variety of demographic and economic characteristics of the U.S. population.   This census series has been made at 10-year intervals since 1790; the last one for which reports are available at the time of this writing is for 1960 (count made as of July 1 for the year). A whole series of individual reports covering individual subject areas derived from the data collected is available.

County Business Patterns

An annual compilation of statistics for each county and each state showing total employment and number of business establishments in each of eight employment size categories, for each kind of business defined in the Standard Industrial Classification system. It is compiled for the employment as of mid March, using Social

Security tax reports, which by law must be filed with the Internal Revenue Service by every employing establishment. It is compiled and published in 53 individual reports representing a U.S. total summary, and a state total plus individual county breakdowns for each state, the District of Columbia, and the outlying territories. Some individual state reports may be available as early as May of the following year, with the complete set usually ready by September. Individual state reports have varying prices, with the full set at $35.20. This is probably one of the best tools available for the researcher to use in estimating size and geographic distribution of industrial and commercial markets—once he has identified the kinds of establishments that can use his product and their probable usage rates.

Bureau of the Census Catalog

Issued quarterly with each issue listing all the material or reports published by the Bureau during the current year to date; thus, the issue published at the end of the fourth quarter is a complete listing for the year. Gives an explanation of each report or series, mentions what parts of a series were published in previous years and what part, if any, is still to come. Supplements are issued monthly covering only material released during that month and without detailed explanatory comment. Each of the individual reports in a census series is listed and explained as it is issued. A researcher should have available the fourth quarter issue of the Catalog for several preceding years to identify all the individual reports of any series that may be of value. The catalog also lists machineable data files (punch cards or tape), special tabulations, and other unpublished material on a cost basis from the Bureau. Catalog is available at varying prices for single issues or at an annual subscription rate of $2.25.

Survey of Current Business

A monthly publication giving current monthly statistical data on some 2500 different series of economic factors and indexes: plus editorial material on selected business conditions and economic research techniques. About every two years a supplementary issue is published giving comprehensive annual historical series for each of the indicators and economic factors that are followed on a monthly basis. This is a useful source of basic information for trend analysis and forecasting correlations. Published by U.S. Department of Commerce, Office of Business Economics, available on single issue basis or annual subscription.

U.S. Government Organization Manual

Published annually by the General Services Administration. This is the official "organization" handbook of the Federal Government. Describes the various branches, departments, agencies and what they do; lists representative kinds of material published.

Monthly Catalog of U.S. Government Publications

Issued monthly by the Superintendent of Documents, U.S. Government Printing Office. This covers every item published by every agency of the federal government during the month. The December issue contains a cumulative index for the whole year but requires reference to the individual monthly issues for descriptive and price details. Descriptions in this catalog are minimal. This is a source for keeping aware of the kinds and amounts of information being generated by agencies of the federal government other than the Bureau of the Census. Annual subscriptions, $4.50; individual issues 50¢ except the December index issue, which varies in price.

Marketing Information Guide

Published monthly by the U.S. Department of Commerce, Business and Defense Services Administration. This is an annotated, descriptive bibliography of currently published secondary statistics and other marketing information; including textbooks, monographs on technique development, and specific market research studies generated by private business and institutional organizations as well as by governmental organizations. It is an excellent means of keeping aware of new or updated information and sources that may be helpful in the marketing function as it becomes available. Cumulative indexes of all items listed previously during the year are included in the March, June, September and December issues. Annual subscription is $2.00.

All these federal government publications that are still in print (including back issues of regular periodical material and final reports of previous census series) can be obtained from the Superintendent of Documents, U.S. Government Printing Office, Washington, D.C. 20402. Order forms describing and pricing all the individual reports available in the three different census series can be obtained on request from the Bureau of Census or the Superintendent of Documents.

All the items are found in university and public libraries that have been designated as government depository facilities; they may also be found in many other libraries. The current Department of Commerce material is generally available in field offices of the Department of Commerce.

Encyclopedia of Associations, Volume I, National Organizations of the U.S.

Published by Gale Research Company, 1400 Book Tower, Detroit, Mich. This is revised and updated about every two or three years. It is a comprehensive listing of associations in the United States, classified and arranged by *areas of interest*, and showing headquarters address and name of administrative director, number of staff personnel, number of members and primary purpose and activities of the association. Last available edition at the time of this writing was 5th Edition, January, 1968—$29.50.

American Paper Institute

260 Madison Avenue, New York, N.Y. 10016

This association is listed specifically because it is specific to the paper industry. It does have extensive programs for compiling paper and paperboard statistics based on the production and shipment records of its individual members. In addition, it has an active program for the improvement of marketing and the promotion of marketing research within the industry. This is supported with staff personnel and staff activities. A good reference library on marketing and marketing research has been assembled at the association headquarters, and it naturally includes considerable source material specific to the end uses and markets for paper or paperboard. Although in general each of the statistical series compiled by the association is available *only* to those members participating in that series, much useful research guidance and summary material relevant to specialty paper can be provided by the staff of this association.

The following three specific association entries are also included in our select list because of their well-known programs of emphasis on marketing research "know-how" and application. Each has published numerous bibliographies or lists of marketing research reference material and/or information sources. These range from general compendiums to lists of specific factors, such as a bibliography on mathematical methods in marketing, or a list of marketing consultants that are members. Rather than attempt to select individual sources published by any association, it is more appropriate to direct the information seeker to the staff and headquarters of each, where he can learn about the total program of the association, take advantage of assistance and guidance in research technique as well as information sources, and obtain the most recent, updated sources of information available.

American Marketing Association
527 Madison Avenue, New York, N.Y.

Note particularly their "Basic Bibliography on Marketing Research" published in 1963; their "Geographic Listing of Marketing Consulting and Research Agencies" which is revised and updated at varying intervals; and their "Journal of Marketing" and "Journal of Marketing Research", quarterly and bimonthly periodicals that publish current information in that field.

American Management Association
135 West 50th St. New York, N.Y.

Note particularly their "Guidelist for Marketing Research and Economic Forecasting" published in 1966, their "Directory of Consultant Members" which is updated biannually, and their special reading lists or bibliographies for individual AMA seminars, especially those in the fields of marketing, product planning, or marketing research.

National Industrial Conference Board
845 Third Avenue, New York, N.Y. 10022

Note particularly their "Studies in Business Policy," which deal with Marketing and Marketing Research problems and applications, their frequent special studies of economics and business forecasts, and certain continuing economic series they compile and publish regularly in the Conference Board Record.

Business Publication Rates and Data

Published monthly by Standard Rate and Data Service, Inc., 5201 Old Orchard Road, Skokie, Ill. 60076. Annual subscription price, $44.00. This directory lists the advertising rates and production information for business and trade magazines that accept paid advertising. It is updated and published monthly for use by advertisers in their selection of media and preparation of trade advertising material. It is, however, one of the most comprehensive listings available of *business and trade magazines* published in the U.S. The magazines are classified and arranged in groups by some 160 different primary *areas of interest,* and the data listed for each magazine include the address of the publication office and the key management personnel. In general, the business and trade magazine publishers are more apt to be sources of marketing information on specific products for industrial or commercial use than are the other communications media; although Standard Rate and Data Service, Inc., does publish listings of Rate and Data information for each of the "advertising" media. In addition, SRDS, Inc., develops and publishes geographic profiles of certain consumer market factors that may be useful for defining or estimating the geographic distribution of market potential. Thus, the SRDS, Inc., organization itself might be worthy of separate mention on our list of selected sources.

Sales Management's Annual Survey of Buying Power
   Published annually in June by Sales Management Magazine, Inc. 630 Third Avenue, New York, N.Y. 10017—Price $6.00. Gives current adjusted estimates of population and family units as of December 31 of preceding year, plus total consumer purchasing power (Disposable Personal Income), total retail sales, and retail sales for each of several broad consumer product classes for the preceding year, broken down by individual state, county and metropolitan area.

The following six publishers of trade magazines are listed because of their recognized editorial activities and interests in paper production and the paper markets:

Peacock Business Press, Inc.
   200 South Prospect Ave., Park Ridge, Ill. 60068
Publications of interest in the paper field include:
   "American Paper Merchant"—L.Q. Yowell, Editor
   "Paper, Film and Foil Convertor"—Vernon A. Prescott, Editor
   "Source of Supply Directory"—L.Q. Yowell, Editor

Ojibway Press, Inc.—Ojibway Building, Duluth, Minn. 55802
   "Modern Convertor"—Don Greuning, Editor
   "Paper Sales"—Roy Wirtzfeld, Editor
   "The Paper Yearbook"—Roy Wirtzfeld, Editor
   "Office Products Dealer"—Tim Donovan, Editor

Lockwood Publishing Co., Inc.—551 Fifth Avenue, New York, N.Y. 10017
   "Lockwood's Directory of the Paper and Allied Trades"—Robert Sanford, Editor
   "Paper Trade Journal"—John C.W. Evans, Editor

Waldon/Mott Corporation—466 Kinderkamack Road, Oradell, N.J. 07649
   "Paper Age"—Robert C. Tate, Editor
   "The Paper Catalog"—Roy Osborn, Jr.—Editor
   "Walden's ABC Guide and Paper Production Yearbook"—Roy Osborn, Jr., Editor
   "Printing Magazine/National Lithographer"—James F. Burnes, Jr., Editor

"Printing Purchasing Manual"—Roy Osborn, Jr., Editor
McGraw-Hill Publications—330 West 42nd Street, New York, N.Y.
10036.
"Chemical Week"—Howard C.E. Johnson, Editor
"Modern Packaging"—Robert J. Kelsey, Editor
"Modern Packaging Encyclopedia"—William C. Simms, Editor
The McGraw-Hill Research Department offers special marketing
research services for surveys in any manufacturing industry.
Magazines for Industry—A subsidiary of Cowles Communications, Inc.
777 Third Avenue, New York, N.Y. 10017
"Paperboard Packaging"—Richard H. Green, Editor
"Food and Drug Packaging"—Marvin Tobin, Editor
"Hardgoods and Soft Goods Packaging"—Max Baxley, Editor
"Official Container Directory"—Fred Sharring, Editor
Other sources are:
Directory of University Research Bureaus and Institutes
Gale Research Co., Inc.—1400 Book Tower, Detroit 26, Mich.
At the time of this writing last available edition—1965
Associated University Bureaus of Business and Economic Research
Bureau of Business Research, University of Oregon, Eugene, Ore.
97403. A membership list of this association identifies prospective
university related sources of research assistance and information.
The association issues an annual bibliography of all books, reports
and monographs published by university research bureaus.
Sources of State Information and State Industrial Directories
Published in 1964 by the U.S. Chamber of Commerce, 1615 H
Street, N.W., Washington, D.C. 20006. This gives names and ad-
dresses of public and private agencies that provide information about
individual states. In addition, it lists the availability and source of
individual state directories of manufacturers. These state industrial
directories are excellent sources for identifying the names and addres-
ses of specific kinds of manufacturers who might be the primary
sources for a particular research project.
*Bradford's Directory of Marketing Research Agencies and Manage-
ment Consultants in the United States and the World.*
Revised and published biannually by Bradford's Directory, P.O.
Box 207, Middleburg, Va. This listing describes size and qualifications of staff, and the types of research services undertaken for
each organization listed.
International Directory of Marketing Research Houses and Services.

Annual directory published by Marketing Review, the house organ of the New York Chapter, Inc., American Marketing Association, 527 Madison Avenue, New York, N.Y. 10022, $4.00.

Directory of Special Libraries and Information Centers

Second edition published by Gale Research Co., 1400 Book Tower, Detroit, Mich., 1968; price, $28.50. This is a guide to collections and personnel in U.S. and Canadian special libraries; information, documentation, and data centers, and archives that are sponsored by governmental agencies, business firms, trade associations, and professional societies. The 11,500 installations are arranged alphabetically by name of supporting organization. For each facility, the entry gives its official name, address, and telephone number; names and titles of professional personnel; subject fields; and services available. A subject index is included.

Guide to American Directories

Seventh edition published by B. Klein and Company, 104 Fifth Ave. New York, N.Y., 1968; Price, $25.00. A listing of some 4500 directories of all kinds published in the U.S., giving descriptive information on each; arranged in some 300 "subject" categories.

Two other miscellaneous sources that rate mention in this selected list of sources because of the frequent help they have provided the author in identifying specific companies involved in a specific type of business are:

Classified Listings or Yellow Pages of local telephone directories
Credit Rating Reference Book Service of Dun and Bradstreet, Inc. 99 Church Street, New York, N.Y. 10007. In addition, the DUN'S data files maintained by the Marketing Service Co., Division of Dun & Bradstreet, Inc., provide specialized listings and information on specific kinds of companies. The Marketing Service Co. Division also provides other specialized marketing research services.

METHODS OF COLLECTING DATA IN MARKETING RESEARCH

All marketing information, whether available from secondary sources or resulting from an original research of primary sources, must involve at some point an original collection of basic data from primary sources. The reader ought to have some familiarity with the methods used to collect original data. These are generally classified into three basic procedures—the *survey* method[57], the *observational* method, and the *experimental* method[22,76]. The researcher faced with collecting original data from primary sources should explore the possibilities of different variations of each method—and sometimes a combination of more than one method is needed for best results.

Survey Method

In the survey method, data are gathered by asking questions of the primary sources. The survey or questionnaire approach, the most widely used of the data collecting methods, is unfortunately sometimes considered synonymous with marketing research. Some researchers use the survey technique when one of the other methods would be more appropriate. Surveys are often classified into three basic types—*factual* surveys, *opinion* surveys, and *interpretive* surveys. The distinction is important, and the application of information developed is usually for different purposes.

Factual Surveys

Factual surveys ask questions about existing facts—What material are you using? What brand? How much did you use last week? When did you make your last purchase?

Opinion Surveys

Opinion surveys ask for judgments, appraisals, evaluations, or opinions of the respondent—Which of these designs do you think is most appealing? What is the most important performance characteristic for this product? Which supplier has the most helpful promotion program? The opinion survey method provides qualitative analyses and understanding of customer attitudes and beliefs for guiding decisions on product design and for the planning of marketing programs. The existence of a definite qualitative public opinion toward a manufacturer, a dealer, or a product is one of the strongest assets or liabilities encountered in marketing.

Interpretive Surveys

Interpretive surveys ask the respondent to explain why he buys a certain product or why a certain attitude exists. This is the most difficult of surveys because respondents often cannot explain why they behave or react in a particular way. Such surveys are often prompted because facts and opinions reported do not explain what is happening in a particular marketing situation. This type of survey usually requires the specialized questioning techniques of the behaviorial scientist, sometimes referred to as motivational research.

Observational Method

The observational method relies upon direct observation of *physical* phenomena in collecting data. For example, the researcher may actually check the brands on pantry shelves or the material in the supplies storeroom of a conpany. Direct observation may be made by mechanical means as well as personal observation. Cameras can record facial expressions and dilation of the pupil of the subject's eye; or electrical counters can record frequency of certain purchasing actions. Generally, this method is more objective and accurate than the survey method but it has the limitation that it can be used only where some physical phenomena or action takes place that can be measured and/or counted directly.

Experimental Method

The experimental method makes use of controlled comparisons of alternative marketing plans, or sales tests of alternative product designs. Variables in the marketing "mix" (product, price, promotion, service, and distribution) are arranged and controlled in two or more actual sales situations so that their influence on the actual sales results can be measured. A common application of this approach is in the use of test markets to note the effects of changes in advertising, distribution, product design, packaging, or promotional appeals. Its use normally occurs in the final phase of a product development or new business development situation to check the actual marketability of possible alternative product designs and/or marketing programs before going into full scale commercial production and investment. The success of this type of research places a premium on knowledge of experimental procedures, and the ability to design and set up properly matched and controlled conditions for the tests.

KINDS OF MARKETING RESEARCH AND APPLICATIONS

Professional marketing researchers are often specialists who concentrate their activity in a particular type of research. Many independent marketing service agencies have been organized to provide specialized information services based on a particular type of research or a particular application of research. The reader should be aware of the various types of marketing research and their normal areas of application to be judicious in his selection of research assistance to meet a particular information need.

On a broad basis, a distinction is generally recognized between *consumer* marketing research and *industrial*[94] marketing research; that is, research undertaken to develop information about the marketing of goods or services to people for personal use, and research undertaken to develop information about the marketing of goods or services to business establishments for use in their production or conduct of business. Sometimes a third broad class is distinguished—*institutional* marketing research undertaken to develop information about the marketing of goods or services to institutional organizations, particularly government operations. Normally, however, this is grouped with industrial marketing research.

Within these broad classes, further distinctions in marketing research are generally identified and classified according to the marketing function or the application for which information is sought. In recent years, however, theorists have added some new "types" based on new sophisticated techniques that have been developed for inquiry or for the analysis of data.

Modern texts and references identify at least fourteen "types" of marketing research. They are briefly:

1. *Product Analysis*

This includes marketing research projects aimed at (a) locating or identifying market needs that might be satisfied with a new kind of product or a modification of an existing product (product concept); (b) determining the specific product design and performance characteristics required to satisfy the market (product design), and (c) measuring how well the product satisfies the design and performance requirements of the market (product acceptance). "Product" in this context includes services as well as physical products. It also includes the packaging and the variations in size and color for an effective "line." This type of research normally requires original data collection at the customer level. It may employ either the survey, the observational or the experimental approach, although the predominant method is survey. It is probably one of the most frequently needed types of marketing research in a specialty papers business.

2. *Customer Analysis and Classification*

This includes research aimed at (a) locating and identifying the actual individual customers or potential customers for a product, (b) determining the distinguishing characteristics of actual customers (e.g. the age, sex, race, income level, geographic location, how and why product is used, and product volume requirements for "consumer" products—or the kind of business, size of business, geographic location, how and why product is used, and product volume requirements for "industrial" products), and (c) classifying the individual customers into specific "market segments" based on common characteristics of end use, kind of customer and geographic location. This type of research can often be accomplished through the use of internal company sales records and secondary directory sources.

3. *Analysis of Total Market Potential*[106]

This includes research aimed at estimating the total number of potential customers as well as the product volume by "market segment." There are different approaches employed for this type of research depending on the particular situation and the amount of secondary information available. They range from a breakdown of total industry production data by using statistics on related economic factors available from secondary sources to a build-up of total potential based on original research surveys that measure the number of potential customers and the volume usage for individual "market segments" within statistically representative samples of the total population.

4. *Market Trend Analysis and Forecasts*

This includes research aimed at (a) measuring changes that take place in numbers of potential customers, in their product design and volume requirements, and in their opinions and attitudes about the products and marketing practices of suppliers, (b) relating these changes to changes in the economic, cultural and/or technological environment, and (c) projecting the effect of these changes on the volume and the nature of the market at future points in time, either short range or long range. This type of research is normally based heavily on secondary sources of information, although specific cases often require supplementary original source research.

5. *Distribution Channel Research*[12,62,107]

This includes research aimed at (a) determination of the number and the kinds of customer contacts, plus other marketing services such as credit, delivery and technical assistance that are required to market the product, (b) determination of what physical distribution facilities and what distribution channels are available to provide these services, (c) determination of the supplier marketing services required for effective functioning or motivation of the various "kinds of dealers" available, (d) selection of individual dealers, and (e) measurement of the performance of individual dealers. This type of research usually requires direct original survey work at both the customer and the dealer level.

6. Research of Marketing Organization and Operations

This research is normally aimed at establishing or modifying the internal organization of company personnel, operating policies and procedures, and compensation programs for more effective marketing of the company's products to the "market segments" it wants to serve. It may be undertaken by either suppliers or dealers, although suppliers may sometimes initiate studies of the organization and operations of their dealers. Such studies, of course, must be made in the light of information on the characteristics and marketing requirements of the customers (Customer Analysis), the number and geographic distribution of customers (Analysis of Market Potential), and the distribution channel requirements (Distribution Channel Research), to identify the critical areas and the priorities for concentration of marketing and sales efforts. Where the study involves existing operations on established products, it also needs information on the company's marketing performance (see the succeeding types of marketing research identified as "Sales Analysis," "Marketing Cost Analysis," "Measurement of Relative Market Position," and "Company Image Research"), to identify the weak spots in the marketing effort. This type of research, particularly where "problems" with existing operations are indicated, normally requires original source investigation by means of the observational method. Modifications of the Industrial Engineering technique generally known as Time and Activity Analysis find frequent application. This type of investigation is probably more often associated with Management Consultants than with conventional Marketing Researchers.

7. Sales Analysis[10]

The evaluation and comparison of a company's volume and profit results on its own sales, broken down in detail by product lines or individual product items, geographic territory, size of order, kind of customer, or by individual customer or dealer, end use, and time periods is a powerful tool for researching strengths and weaknesses in the company's marketing programs and operations. It can point up growth

and profit opportunities that might be exploited—or problem areas
that should be corrected or discontinued. It can help identify and
define specific marketing research projects of other types that should
be undertaken. This type of research concentrates on the analysis
phase of the research procedure rather than the data collection phase;
but its effectiveness is directly related to the amount of detail on individual sales orders that is organized and retained in the company records.
A major part of this research can be effectively programmed for computer collection, storage, and analysis of data already entered in
electronic data processing systems for the processing of individual sales
orders and invoices.

8. *Marketing Cost Analysis*[34,59,108,109]

This is research aimed eventually at identifying unit costs of every
marketing and distribution operation involved in moving goods from
the production line to the customer in such terms that the true portion
of the costs can be assigned to each product item of each sales order
filled. This type of research is called "Distribution Cost Analysis" by
many writers and authorities. It uses the analysis techniques of cost
accounting, and the observational or experimental methods of data
collection.

9. *Pricing Analysis and Research*[54,110,111,112]

Included in this category is research aimed at (a) discovery of the
relationship of the price to product quality and service in the customers'
value concepts for a particular product, (b) determination of the actual
finite range of prices that would be competitively acceptable, and (c)
measurment of the effect of different prices within this range on potential
sales volumes and profits. The pricing of products is a major field of
economic study in itself, a considerable amount of literature is available
on the theory and practice of pricing for both consumer and industrial
products. Establishing and maintaining a product/service/Price package
that competitively satisfies the customer's value requirements and yields
á *profitable* sales volume is vital for successful business on any product
at any time. It is particularly important for the successful marketing of
new products. Thus, it is a prime concern of the specialty papers business, which is so dependent on the development and introduction of new
products. Research of this type should be undertaken early in the
concept stage of a new product to establish design targets of price and
cost as well as physical characteristics and performance. Pricing for a

specific product/service package almost always requires investigation at the original source level. In the concept phase of a new product such investigation may have to consist of the survey method; however, the experimental method with test marketing techniques is by far the most accurate and effective for decisions on final pricing.

10. *Measurement of Relative Market Position*

Evaluation of a company's own percentage of the total market is included in this type of research. This, of course, is a fairly simple ratio if total market potential has been determined. Where it is important to have information on the relative shares of all the major competitors, and on trends or changes in those relative shares, the research problem becomes considerably more complicated because it is without the benefit of the sales records of the competitive companies. Original data on the purchase or use of each competitive product must then by collected by survey or observational methods among a representative sample of the market. This type of marketing research is used very extensively in the tactical planning of marketing programs for highly competitive consumer products. Many specialized syndicated research businesses have been established to provide this kind of information on relative market position of selected products for groups of subscriber clients.

11. *Company Image Research*

Students of marketing know that beliefs and attitudes about a company that are consistent or generally held among its potential customers are major factors affecting its sales performance. Well-managed companies make frequent use of this type of research to keep abreast of the current opinions of the marketplace about them and about their competitors. They try to maintain a clear understanding of how the potential customers "view" the "characters" of the companies in general, the quality and reliability of their products, and the adequacy of their marketing programs. Then they plan their marketing and promotion programs to emphasize the favorable factors in their own "image," and to correct or change the less favorable. This is a fairly specialized type of research requiring direct survey of the original sources whose opinions and attitudes are to be measured. Special attention is required in design of questionnaire and in conduct of interviews to avoid influencing the answers given in any way. Adequately large and accurately representative samples are important for meaningful results.

12. *Advertising and Promotion Research*[7,9,15,115,132]

This includes any marketing research undertaken specifically to provide a basis for the development of advertising or promotion programs, or to evaluate their effectiveness. Many of the other types of research are directly adaptable to this application. Knowledge of who and where the potential customers are is necessary, but also one must know the physical means that will deliver a message to them and at what cost. One must know what the customers' product requirements are and what they consider the most important features and also why they buy and how they make the purchase decision (product and service appeals). Knowledge of the attitudes and opinions of the customers concerning the advertiser and his products is important, but some understanding of their personal interests and motivations (psychological appeals) is also needed. Much specialized research for measuring the cost and effectiveness of the various kinds of communications media and for the discovery and evaluation of various "copy" appeals has evolved. *Media* research may use a considerable amount of secondary source information. *Copy* research almost always requires direct investigation at original source levels. Any one, or a combination of the data collection methods of survey, observation, or experiment may prove to be the most effective for a particular project.

13. *Behavior Research*[6,25,116,117,118]

This is one of the newer types of marketing research adapting the concepts and the research techniques of the behavioral sciences to a discovery and explanation of the factors of human behavior patterns that affect marketing situations. Conventional marketing research techniques had been able to determine *who* the customers were, *what* they bought and *how* they bought, and it could measure their purchasing reactions to changes in marketing programs, but it was all too often stymied in attempts to answer *why*. Individuals are seldom conscious or perceptive of basic personal and social factors influencing their purchasing decisions, and are thus unable to give accurate answers to direct "why" questions. Then psychologists and psychoanalysts came along with some insights on the "why" questions; in the early 50's, enthusiasts and "promoters" hailed "Motivation Research" as *the* method to use for all marketing studies. Over-promotion and wide misuse gave motivation research an unfavorable reputation in many quarters, but wise researchers and business managers recognized it as an alternative research approach that was more appropriate than other methods for developing certain kinds of information. They used it in combination with other methods as a powerful tool for formulating hypotheses about specific marketing situations that were then proved or disproved by conventional research. They went on to adapt the concepts and the research techniques of other behavioral sciences, particularly anthropology and sociology; and they recognized that "motivation" research was an unfortunate misnomer since many factors motivating the market had been researched effectively by means of conventional methods long before the techniques of the behavioral sciences were adapted. The techniques of behavioral research require the specialized training and expertise of professional behavioral scientists for proper application and interpretation in the marketing situation, just as they do in any other field in which they are used.

14. *Operations Research/Market Simulation*[39,77,119,120,121]

This technique or type of research involves the construction of a set of mathematical formulas or a "model" representing the relationships and interactions of all the factors in a particular "operation" or process. The technique was engineered to a level of practical application by scientists and mathematical theorists during World War II for solving problems in military operations and development of systems of weapons. Several more or less standard "models" or theoretical sets of relationships have been developed to represent different kinds of operations or systems. In recent years many of these have been effectively adapted to the analysis of individual marketing functions or of total marketing situations (Market Simulation). Some of the better known "models" with typical marketing applications are:

A. *Linear Programming*

Studies of physical distribution problems; salesmen coverage of territory

B. *Queueing or Waiting Line Theory/Monte Carlo Simulation*

Studies of inventory control vs. customer service requirements

C. *PERT/CPM—Program Evaluation Review Technique/Critical Path Method*

Planning of the scheduling of events and the allocation of resources for new product development and introduction or for advertising campaigns

D. *Games Theory*

Identification and analysis of alternative possibilities for competitive marketing strategies and tactical manuevering

E. *Markov Process*

Analysis of and model simulation of "brand" switching, particularly in consumer marketing situations

F. *Bayesian Decision Tree*

This is a technique for structuring a model of the branching of alternative events or results from alternative decisions, assigning *probability judgments* to the alternative results, and sequential analysis for "best" decisions on the basis of these probability judgments. It has found effective use in development of new projects for guiding decisions on whether or not to invest in various phases of market evaluation of product concept and design or for test marketing prior to general market introduction.

G. *Dynamic Programming/Stochastic Programming*

Mathematical representation of the interactions of all the functions in a total business system

The "structuring" of a mathematical model for a particular system allows programming for computer analysis, for examining larger numbers of alternatives or processing larger volumes of data within this relationship than would be physically or economically possible without the computer, and for asking "what if" questions and studying their effects through a "simulation" model that represents the real world without committing the capital and manpower necessary for testing them in the real world. Often just the exercise of formally structuring and recording the mathematical relationships provides the insights and knowledge for making sound decisions without any necessity of computer programming. The development of a meaningful and realistic mathematical model of a marketing situation must proceed from factual information about that situation. If such information is not already available, it must be gathered by conventional marketing research methods. The disciplined structuring of a mathematical model is a most effective means for identifying and defining specific marketing information gaps that should be researched to gain a clear understanding of the "system" and to make sound decisions in relation to it. Effective use of this type of research requires the guidance and help of the professional specialist with training and experience in the application of the mathematical techniques involved.

Input-Output Analysis is a technique that is currently receiving a great deal of developmental effort[123-131,incl]. It may well be the next generally recognized addition to this list of kinds of marketing research. It is a technique for identifying and quantifying all of the "inputs" to a particular operation or system (i.e., materials, energy, capital, and manpower) required to provide or "balance" a given level of "output" or final demand on the system. The concept has been around for a long time in the form of energy and/or material balances in the scientific fields of chemistry and physics.

The current interest in developing the technique for economic analysis and marketing applications stems from studies initiated in the 1930's by Wassily W. Leontief, the Henry Lee Professor of Economics at Harvard University. His original efforts were aimed at breaking down and quantifying the interrelationship and the interdependence of the various producing and consuming sectors making up total national economies. During the 1940's, his work was given strong sponsorship and support by various agencies of the federal government in their search for better tools to guide the allocation of resources and the support of industrial activities necessary to meet the economic demands of the war or the total national economy. In the 1950's, the federal agencies, particularly the U.S. Department of Commerce, Office of Business Economics, further supplemented and expanded his concepts with internal development efforts. In 1964 a matrix model of the U.S. economy based on 1958 census data was published by the Office of Business Economics. This broke down the Gross National Product of the United States for 1958, and traced it as dollar values of "inputs" and "outputs" among more than 80 "producing industries" and final demand sectors making up the total economic structure of the country.

The government agencies are continuing their development efforts along the lines of a more detailed breakdown of the matrix model into more than 300 producing and final demand sectors and in evaluation of changes in the interrelationships of inputs and outputs among the various sectors that have occurred between the 1958, the 1963, and the 1967 censuses. In the meantime private commercial and institutional organizations are climbing on the bandwagon for developing and a-dapting this concept in the areas of economic forecasting, business expansion, diversification planning, manpower planning, marketing

planning, and financial control. Extensive sources of secondary information that facilitate further adaptation and application for individual
SIC 4, 5, or 7 digit industries to individual company operations are
being built up.

After this brief review, it should be evident that considerable overlap—
sometimes almost an interdependence—exists between many of the
types of marketing research that have been classified and defined. Any
particular management problem or decision need for marketing information might well use a combination of two or more of these types of
marketing research for most effective results.

BIBLIOGRAPHY

Items 1 through 28 are all published by the National Industrial Conference Board,
Inc., 845 Third Avenue, New York, N.Y.

1. "New Product Development-I, Selection, Coordination, Financing," Studies in Business Policy No. 40, 1950.
2. "New Product Development-II, Research and Engineering," Studies in Business Policy No. 57, 1952.
3. "New Product Development-III, Marketing New Products," Studies in Business Policy No. 69, 1954.
4. "Marketing, Business and Commercial Research in Industry," Studies in Business Policy No. 72, 1955.
5. "Marketing Research in Action," Studies in Business Policy No. 84, 1957.
6. "Use of Motivation Research in Marketing," Studies in Business Policy No. 97, 1960.
7. "Measuring Advertising Results," Studies in Business Policy No. 102, 1962.
8. "Forecasting Sales," Studies in Business Policy, No. 106, 1963.
9. "Pretesting Advertising," Studies in Business Policy, No. 109, 1963.
10. "Sales Analysis," Studies in Business Policy No. 113, 1965.
11. "Measuring Salesmen's Performance," Studies in Business Policy No. 114, 1965.
12. "Selecting and Evaluating Distributors," Studies in Business Policy, No. 116, 1965.
13. "Setting Advertising Objectives," Studies in Business Policy, No. 118, 1966.
14. "Using Marketing Consultants and Research Agencies," Studies in Business Policy No. 120, 1966.
15. "Evaluating Media," Studies in Business Policy No. 121, 1966.
16. "Appraising the Market for New Industrial Products," Studies in Business Policy No. 123, 1967.
17. "The Development of Marketing Objectives and Plans," Experiences in Marketing Management No. 3, 1963.
18. "Building a Sound Distributor Organization," Experiences in Marketing Management No. 6, 1964.
19. "A Philosophy of Marketing for Management," Experiences in Marketing

Management No. 7, 1965.
20. "Emerging Trends in Marketing," Experiences in Marketing Management No. 9, 1966.
21. "Organization for New Product Development," Experiences in Marketing Management No. 11, 1966.
22. "Market Testing Consumer Products," Experiences in Marketing Management No. 12, 1967.
23. "The Marketing Executive Looks Ahead," Experiences in Marketing Management No. 13, 1967.
24. "Why New Products Fail," The Conference Board Record, Vol. I No. 10, October, 1964, pp. 11-18.
25. "What is Behaviorial Science?" The Conference Board Record, Vol. II No. 9, September, 1965, pp. 35-41.
26. "Media Research: A Progress Report," The Conference Board Record, Vol. IV No. 5, May, 1967, pp. 36-40.
27. "What Marketing Executives Want—New Products," The Conference Board Record, Vol. IV No. 5, May, 1967, pp. 17-22.
28. "New Product Risk in Industrial Markets," The Conference Board Record, Vol. IV, No. 11, November, 1967 pp. 26-31.

Items 29 through 64 are all published by the Small Business Administration Washington, D.C., an agency of the U.S. Government. Items 29 through 45 are available free of charge from the SBA Headquarters or Regional Offices only; items 46 through 64 are available for purchase only through the Superintendent of Documents, U.S. Government Printing pffice, Washington D.C.

29. "New Products Development and Sale," Small Business Bibliographies No. 4.
30. "Marketing Research Procedures," Small Business Bibliographies No. 9.
31. "Statistics and Maps for National Market Analysis," Small Business Bibliographies No. 12.
32. "National Directories for Use in Marketing," Small Business Bibliographies No. 13.
33. "Basic Library Reference Sources," Small Business Bibliographies No. 18.
34. "Distribution Cost Analysis," Small Business Bibliographies No. 34.
35. "Market Research and Planning for Small Manufacturers," Management Research Summaries No. 19.
36. "Forecasting in Small Business Planning," Management Research Summaries No. 23.
37. "Use of External Assistance by Small Manufacturers," Management Research Summaries No. 30.
38. "Use of Outside Information in Small Firms," Management Research Summaries No. 56.
39. "Operations Research in Small Business," Management Research Summaries No. 141.
40. "Helping Small Firms Develop and Exploit New Products," Management Research Summaries No. 181.
41. "Using Census Data in Small Plant Marketing," Management Aids No. 187.
42. "Developing a List of Prospects," Management Aids No. 188.
43. "Measuring the Performance of Salesmen," Management Aids No. 190.

44. "Profile your Customers to Expand Industrial Sales," Management Aids, No. 192.
45. "Marketing Planning Guidelines," Management Aids No. 194.
46. "Making Your Sales Figures Talk," Small Business Management Series No. 8.
47. "New Product Introduction for Small Business Owners," Small Business Management Series No. 17.
48. "Technology and Your New Products," Small Business Management Series No. 19.
49. "Practical Business Use of Government Statistics," Small Business Management Series No. 22.
50. "How the Small Plant Can Analyze Old and New Markets," Management Aids for Small Business: Annual No. 1, Chap. 8.
51. "Appraise Your Competitive Position to Improve Company Planning," Management Aids Annual No. 2, Chap. 6.
52. "Sales Forecasting for Small Business," Management Aids Annual No. 2, Chap. 8.
53. "How Trade Associations Help Small Business," Management Aids Annual No. 2, Chap. 17.
54. "How to Price a New Product," Management Aids Annual No. 3, Chap. 2.
55. "How to Set up Sales Territories," Management Aids Annual No. 3, Chap. 3.
56. "How Marketing Research Helps Small Business," Management Aids Annual No. 3, Chap. 8.
57. "Making a Marketing Survey," Management Aids Annual No. 4, Chap. 4.
58. "Reducing the Risks in Product Development," Management Aids Annual No. 5, Chap. 3.
59. "Analyzing Your Cost of Marketing," Management Aids Annual No. 5, Chap. 6.
60. "Wishing Won't Get Profitable New Products," Management Aids Annual No. 6, Chap. 3.
61. "How Business Publications Help Small Business," Management Aids Annual No. 6, Chap. 10.
62. "Checking Your Marketing Channels," Management Aids Annual No. 9, Chap. 6.
63. "Selecting Marketing Research Services," Management Aids Annual No. 9, Chap. 10.
64. "Getting Facts for Better Sales Decisions," Management Aids Annual No. 12, Chap. 4.

Items 65 through 77 are all published by the American Marketing Association, 230 N. Michigan Ave., Chicago, Ill.

65. "A Basic Bibliography on Industrial Marketing," 1958.
66. "Marketing Definitions: A Glossary of Marketing Terms," 1960.
67. "Current Sources of Information for Market Research," 1960.
68. "Report of the Definitions Committee of the American Marketing Association," 1961.
69. "A Basic Bibliography on Mathematical Methods in Marketing," 1962.
70. "Current Sources of Marketing Information," 1960.
71. "A Bibliography on New Product Planning," revised, 1966.

72. "A Basic Bibliography on Marketing Research," 1963.
73. "A Survey of Marketing Research," 1963.
74. "Industrial Directories," 1963.
75. "Marketing Management Annotated Bibliography," 1963.
76. "A Basic Bibliography on Experiments in Marketing," 1967.
77. "A Selected Annotated Bibliography on Quantitative Methods in Marketing," 1968.

Selected Textbooks and Reference Books-

Marketing, General.

78. *Industrial Marketing*, R.S. Alexander, J.S. Cross and R.M. Cunningham; Richard D. Irwin, Inc., Homewood, Ill., (1961)
79. *Principles of Marketing*, R.D. Tousley, E. Clark and F.D. Clark, The Macmillan Co., New York, N.Y., (1962)
80. *Marketing Management: Analysis and Planning*, John A. Howard, Richard D. Irwin, Inc., Homewood, Ill. (1963)
81. *Fundamentals of Marketing*, Wm. J. Stanton; McGraw-Hill Book Co., New York, N.Y., (1964)
82. *Introduction to Marketing Management*, Fred M. Jones, Appleton-Century-Crofts, Inc., New York, N.Y., (1964)
83. *Marketing-Principles and Methods*, C.F. Phillips and D.J. Duncan, Richard D. Irwin, Inc., Homewood, Ill., (1964)
84. *Marketing Handbook*, edited by Albert Wesley Frey, The Ronald Press Co., New York, N.Y., (1965)

Marketing Research, General-

85. *Marketing and Distribution Research*, L.O. Brown, The Ronald Press Co., New York, N.Y., (1955)
86. *Marketing Research Pays Off-Forty Case Histories of Profitable Consumer and Industrial Marketing Research*, edited by H. Brenner, Printers' Ink Books, Pleasantville, N.Y., (1955)
87. *Marketing Research*, Richard D. Crisp, McGraw-Hill Book Co., New York, N.Y., (1957)
88. *Marketing Research-Principles and Readings*, P.M. Holmes, Southwestern Publishing Co., Inc., Cincinnati, Ohio, (1960)
89. *Research Methods in Economics and Business*, R. Ferber and P.J. Verdoorn, The MacMillan Co., New York, N.Y., (1962)
90. *Marketing and Business Research*, M.S. Heidingsfield and F.H. Eby, Holt, Rinehart & Winston, Inc., New York, N.Y., (1962)
91. *Marketing Research*, H.W. Boyd, Jr. and R. Westfall, Richard D. Irwin, Inc., Homewood, Ill., (1964)
92. *Marketing Research*, R. Ferber, D.F. Blankertz and S. Hollander, Jr., The Ronald Press Co., New York, N.Y., (1964)
93. *The Strategy of Marketing Research*, C.R. Wasson; Appleton-Century-Ccrofts, Inc., New York, N.Y., (1964)
94. *Industrial Marketing Research: Management Technique*, N.H. Stacey and A. Wilson; Humanities Press, Inc., New York, N.Y., (1964)

Sampling Theory and Application-

95. *Sampling Techniques*, W.G. Cockran, John Wiley & Sons, Inc., New York N.Y., (1953)
96. *Sample Survey Methods and Theory*, M.H. Hansen, W. N. Hurwitz and W.G. Madow, John Wiley & Sons, Inc., New York, N.Y., (1953)
97. *Sample Design in Business Research*, E.W. Deming, John Wiley & Sons, Inc., New York, N.Y., (1960)

Statistical Analysis Techniques-

98. *Statistical Techniques in Market Research*, Robert Ferber, McGraw-Hill Book Co., New York, N.Y., (1947)
99. *Elementary Statistics*, Paul G. Hoel, John Wiley & Sons, Inc., New York, N.Y., (1960)
100. *Quantitative Techniques in Marketing Analysis: Text and Readings*, edited R. Frank, A. Kuehn and W. Massey, Richard D. Irwin, Jr., Inc., Homewood, Ill., (1962)
101. *Introduction to Mathematical Statistics*, by Paul G. Hoel; John Wiley & Sons, Inc., New York, N.Y., (1962)
102. *Statistical Analysis*, S.B. Richmond; The Ronald Press Co., New York, N.Y., (1964)
103. *Basic Statistics for Business and Economics*, edited by D.A. Leabo; Richard D. Irwin, Inc., Homewood, Ill., (1964)

Product Analysis and Development-

104. *Management of New Products*, Booz, Allen & Hamilton, Inc. New York, N.Y., (1960)
105. *New Product Development*, J. Mahare and D. Coddington, Denver Research Institute, University of Denver, Denver, Col., (1961)

Market Potentials-

106. *Market and Sales Potentials*, Francis E. Hummel, The Ronald Press Co., New York, N.Y., (1961)

Distribution Channels-

107. *Distribution Channels for Industrial Goods*, Wm. M. Diamond, Bureau of Business Research, Ohio State University, Columbus, Ohio, (1963)

Distribution Costs-

108. *Distribution Costs*, J.B. Heckert and R. Miner, The Ronald Press Co., New York, N.Y., (1953)
109. *Practical Distribution Cost Analysis*, D.R. Longman and M. Schiff, Richard D. Irwin, Inc., Homewood, Ill., (1955)

Pricing-

110. *The Theory of Price*, G. Stigler, The Macmillan Co., New York, N.Y., (1947)
111. *Price Practices and Price Policies-Selected Writings*, Jules Bachman, The Ronald Press Co., New York, N.Y., (1953)
112. *Pricing for Marketing Executives*, A.R. Oxenfeldt, Wadsworth Publishing Co., Inc., Belmont, Calif., (1961)

Advertising and Promotion Research-

113. *Criteria for Advertising and Marketing Research*, Advertising Research Foundation, New York, N.Y., (1953)
114. *How to Increase Advertising Effectiveness*, Richard D. Crisp, McGraw-Hill Book Co., New York, N.Y., (1958)
115. *Measuring Advertising Effectiveness*, D. Lucas and S. Britt, McGraw-Hill Book Co., New York, N.Y., (1963)

Behavioral Research-

116. *Marketing and the Behavioral Sciences*, edited by Perry Bliss; Allyn & Bacon, Inc., Boston, Mass., (1963)
117. *Modern Marketing Research; a Behavioral Science Approach*, F.T. Scheier, Wadsworth Publishing Co., Inc., Belmont, Calif., (1963)
118. *The Behavioral Sciences Today*, Basic Books, Inc., New York, N.Y., (1963) Operations Research and Mathematical Model Techniques-
119. *Applied Statistical Decision Theory*, H. Raiffa and R. Schlaifer, Harvard University Graduate School of Business Administration, Boston, Mass., (1961)
120. *Mathematical Models and Methods in Marketing*, by F.M. Bass, Richard D. Irwin, Inc., Homewood, Ill., (1961)
121. *Network Analysis for Planning and Scheduling*, A. Battersby, St. Martins Press, Inc., New York, N.Y., (1964)
122. *Econometric Theory*, A.S. Goldberger, John Wiley & Sons, Inc., New York, N.Y., (1964)

Miscellaneous Reports & Magazine Articles-

123. "The Interindustry Structure of the United States," original article on development of Input/Output tables for the National Economy published in survey of Current Business, U.S. Dept. of Commerce, Office of Business Economics, Washington, D.C., November, 1964.
124. "The Structure of the U.S. Economy," (input/output relationships) Scientific American, April, 1965; Scientific American, Inc., New York, N.Y.
125. "Toy or Tool," (Input/Output Analysis), Forbes, Forbes, Inc., New York, N.Y. June 15, 1967.
126. "Input-Output's New Versatility in Sales and Marketing"; Sales Management August 15, 1967; Sales Management, Inc., New York, N.Y.
127. "Planners Put Big Picture on a Grid," Business Week, September 23, 1967, McGraw-Hill Publications, New York, N.Y.
128. "Input-Output Analysis: No Longer a Laboratory Oddity," Sales Management, October 1, 1968., Sales Management, Inc., New York, N.Y.
129. "Marketing Uses Input-Output Analysis," Chemical & Engineering News, September 18, 1967, American Chemical Society, Washington, D.C.
130. "Input-Output Analysis: A Tool for Management and Research," Battelle Technical Review, April, 1968, Battelle Memorial Institute, Columbus, Ohio.
131. "Input-Output Searches The Seventies," Sales Management, August 1, 1968, Sales Management, Inc., New York, N.Y.
132. "Simmons Study of Selected Markets and the Media Reaching Them," W.R. Simmons & Associates, New York, N.Y. 1963.
133. "Look Magazine National Automobile and Tire Survey," Alfred Politz Re-

search, Inc., for Cowles Communications, Inc., New York, N.Y. 1964.

134. "Guidelist for Marketing Research and Economic Forecasting," an information source study prepared for American Management Association, Inc., New York, N.Y., 1966.

135. "Measuring Markets, A Guide to the Use of Federal and State Statistical Data," U.S. Dept. of Commerce, Business and Defense Services Administration, Washington, D.C., 1966.

136. "Standard Industrial Classification for Effective Marketing Analysis," Marketing Science Institute, Philadelphia, Pa., 1967.

chapter 3

Spunbonded Products

ROBERT A.A. HENTSCHEL

INTRODUCTION

Research in the laboratories of E. I. du Pont de Nemours & Co., Inc. has developed a new group of. fibrous sheet structures described as spunbonded products. Generically, spunbonded products are defined as continuous-filament fibrous structures that can be made in the form of fabrics, sheets, and tapes, and are prepared from synthetic polymers in a process integrated with fiber manufacture. They are differentiated from conventional nonwovens in being made from continuous fibers by an integrated process and in having generally higher levels of physical properties. The processes used to make spunbonded products are not related to any of the paper-making processes, and the products are not "papers", although certain of the spunbondeds do have the stiffness, opacity, and smoothness characteristics ordinarily associated with papers. Others of the spunbondeds depart widely from this appearance and are more closely related to fabrics in softness and drape.

A discussion of these products is included in this volume because spunbondeds can, in many cases, be converted on the same equipment and by the same processes as are used in converting papers. Thus, they provide a new group of high-strength, high-performance substrates that can be used by converters of paper to supplement and extend their product lines. In many cases, such products can serve markets that are not open to paper-based materials.

Currently, three general types of spunbonded products are being offered by Du Pont. These include three chemical types and two structural types. The chemical types include one made from a poly-

81

ester, one from polypropylene, and one from linear polyethylene. For the polyester and polypropylene types, the configuration consists of a web of continuous filaments of generally uniform circular or trilobal cross section, the fibers being similar to normal synthetic textile fibers. The nature of the linear polyethylene product, on the other hand, differs in that it is made up of very fine fibers that are interconnected in a continuous network structure.

Table 3-1 gives a general comparison of these three types of spunbonded structures, showing the ranges in individual properties available in each, and how they relate to one another. Although this table covers ranges for individual properties, combinations of these properties can not be freely chosen because when one property is fixed it influences the possible values of others. Each of the spunbondeds is made in a number of styles; the specific properties of each of the various styles is found in later sections of this chapter.

TABLE 3-1
Ranges of Individual Properties
Spunbonded Sheets

Chemical Type	Polyethylene	Polyester	Polypropylene
Du Pont Trademark	Tyvek®	Reemay®	Typar®
Melting Point, °C	132	250	170
Weight, oz/yd²	1.0–2.7	1.3–6.0	2.5–3.5
Thickness, mils	4–9	10–30	9.5–13
Strip Tensile, lb/in	10.5–63	6–54	27–32
Strip Elongation, %	27–36	47–107	33–38
Grab Tensile, lb	25–100	20–160	85–144
Grab Elongation, %	20–30	52–96	25–42
Tear, Elmendorf, lb	0.8–4.5	1.8–13.0	6.0–12.0
Tear, Tongue, lb	2.5–4.4	2.5–12.0	9.0–17.0
Frazier Air Permeability ft³/(ft²)(min)	<1–20	120–813	68–111
Mullen Burst, lb/sq in	47–225	25–140	165–220
Opacity	high	low	low
Recovery & Resilience	low	high	intermediate

The spectrum of properties and characteristics available in spunbonded sheets has led to their application in a large variety of end uses. For the most part, these uses are in the nature of reinforcements, and are either hidden or industrial in character. For example, Tyvek® spunbonded polyethylene is used as a coating substrate for bookbindings, wallcoverings, educational maps, charts, in printed form for tags, labels, advertising signs and banners, for limited-use industrial

protective clothing, and a variety of heavy-duty packagings. Reemay®
spunbonded polyester is used for garment interlinings, reinforcements
in shoe construction, electrical insulations, vinyl sheeting, and in disposable consumer apparel. Typar® spunbonded polypropylene was
developed and is used primarily as the backing or strength member in
the manufacture of tufted carpets. As with the nonwovens, the esthe-
tics of spunbonded sheets do not lend themselves to exposed apparel
uses except for a few special cases.

In the following sections of this chapter, the specific characteristics,
properties, conversion and applications of each of the three spunbonded
types are discussed separately in greater detail. As these products are
still quite new, all having been brought into commerical production
since 1964, further modifications will be made available as the spunbondeds establish their place in the spectrum of available products.

SHEET STRUCTURE OF
TYVEK® SPUNBONDED LINEAR POLYETHYLENE

Tyvek® spunbonded polyethylene is unique among the commercial
fibrous sheet products in that the basic structural element is a continuous web of interconnected fibers. Figure 3-1 is a photograph of

Fig. 3-1. Fiber network.

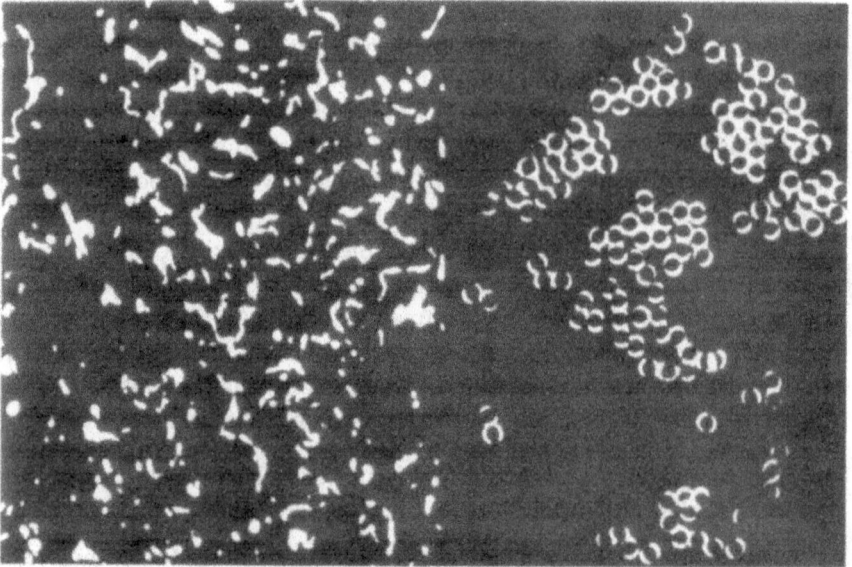

Fig. 3-2. Spunbonded 3 DPF nylon — 100x cross sections.

such a web element, approximately four inches in width. The individual fibers in the network are molecularly oriented. They are randomly variable in thickness from fiber-to-fiber, ranging from less than 0.5 micron to about 10 microns at the extremes and averaging less than 4 microns. Figure 3-2 shows the nature of the cross section. The fineness of the fibers can be judged by comparison with the cross section of a bundle of 3-denier nylon textile yarn filaments, having a diameter of about 20 microns. The web has essentialy no free fiber ends; the fibers interconnect with each other at the points that are, on the average, about 1.5 cm apart. The spunbonded polyethylene sheets are thus built up from very fine fibers arranged in an essentially random pattern. The very fine fibers impart high opacity and whiteness to the sheet, without the use of pigments, because of the scattering of light from the fibers themselves. The random arrangement leads to little directionality of properties.

Spunbonded polyethylene sheets are self-bonded; no external binder is used. By varying the bonding pattern, a considerable variation in sheet properties can be achieved. If the sheets are bonded over their entire area, the resulting products are dense, relatively stiff and smooth, superficially resembling paper. If, on the other hand, the sheets are bonded only at discrete points over the sheet surface, a much softer, more flexible product results. In such point-bonded sheets, further variations are possible through modification of the arrangement and area of the bond points, including types in which the sheet is perforated at the bonds. Because these two product types, area-bonded and point-bonded, are quite different in nature, their characteristics are discussed separately.

AREA-BONDED SHEETS

Physical Characteristics

Area-bonded spunbonded polyethylene sheets are made in a range of weights from 1.3 to 2.7 oz/yd^2. The average physical properties of specific styles in this weight range are shown in Table 3-2. Table 3-3 shows the physical properties of a typical sheet at 2.2 oz/yd^2 together

TABLE 3-2
Average Physical Properties
Area-Bonded Spunbonded Polyethylene

Style No.	1043	1058	1073	1085	Test Reference
Basis Wt., oz./yd.^{2}	1.3	1.6	2.2	2.7	T–410–OS–61
Thickness, mils	5	6	8	9	T–441–M–44
Strip Tensile, lb/in	25/25	33/32	47/46	57/55	T–404–M–50
Strip Elongation, %	22/33	23/28	25/31	27/33	T–404–M–50
Work-to-Break in-lb/in^2	2.5/3.8	4.9/5.1	8.2/9.4	8.9/11.2	T–404–M–50
Tear, Elmendorf, lb.	0.8/0.8	0.9/0.9	1.3/1.3	1.4/1.4	T–414–M–49
Gurley Air Perm., Sec./100 cc in^2	5	7	13	17	T–191B– Method 5452
Mullen burst, lb/in^2	80	140	175	225	T–403–M–53
Spencer Puncture. in–lb/in^2	29	36	60	72	Thwing Manual
Opacity, B&L, %	88	91	94.5	95	T–425–M–60
MVTR, g/(24 hr) (m^2)	965	900	850	700	T–448–OS–49
MIT Flex, cycles	>100M	>100M	>100M	>100M	T–424
Stoll Abrasion, cycles to pinhole	40	63	123	233	CCC–T–191B** Method 5302.1

T = Tappi Method
** = Federal Spec.

Fig. 3-3.   Opacity of sheet structures.

TABLE 3-3
Comparative Physical Properties

	Spun PE Sheet	Kraft Paper	Cotton Sheeting 64×64	Nylon/Rayon Nonwoven
Basis Wt., oz./yd.^{2}	2.2	2.4	2.7	2.4
Thickness, mils	8.0	4.9	10.2	10.1
Strip Tensile, lb/in	46.0	27.9	12.1	7.1
Strip Elongation, %	28.0	4.3	12.0	72.0
Work-to-Break in-lb/in^2	8.0	0.44	1.7	3.0
Tear, Elmendorf, lb	1.3	0.29	2.9	1.2
Mullen burst, lb/in^2	115	25.2	83.4	35.3
Opacity, B&L, %	0.95	0.96	0.63	0.44
CSIA Abrasion, cycles	59	23	18	7
Smoothness, μ in	80	75	280	280
Init. Mod. lb/in.^{2}, ×10^{-3}	120	141	6.5	1.2

All properties are averages of MD and XD values

with the properties of a number of other common sheet structures: kraft paper, a woven cotton sheeting, and a nonwoven fabric of nylon and rayon. All are at approximately equal weights, for comparison. Area-bonded sheets are characterized by a combination of high tenacity and high elongation, resulting in high toughness as measured by the work-to-break. As a result, relatively light-weight sheets can be used to meet end-use demands, especially because the high level of toughness is combined with high opacity. Figure 3-3 illustrates this opacity with the aid of a standard Morest Hiding Power chart, for the sheets described in Table 3-3.

Solvent Resistance

Spunbonded polyethylene sheets are resistant to most common solvents and chemical reagents. They are swelled by hydrocarbons, but are much less affected by polar solvents. Table 3-4 shows the dimensional changes resulting from immersion in a variety of polar and nonpolar solvents. The swelling noted is reversible; the sheet returns to its original size when the solvent evaporates. Physical properties of the sheet are unchanged either in the swollen state or after evaporation of the solvent. These characteristics are important in the formulation of solvent-based coatings and of printing inks for use on spunbonded polyethylene.

TABLE 3-4
**Dimensional Changes Produced
By Solvents on Spunbonded Polyethylene**

Solvent	% Increase In Sheet Length
Water	0.00
Methyl alcohol	0.00
Ethyl alcohol	0.00
Propyl alcohol	0.00
Isopropyl alcohol	0.00
Acetone	0.16
"Cellosolve"[1]	0.31
Methyl ethyl ketone	0.39
Butyl alcohol	0.47
Amyl alcohol	0.47
Ethyl acetate	0.47
Methylene chloride	0.52
Methyl isobutyl ketone	0.78
Hexane	0.94
Cyclohexane	1.10
Tetrahydrofuran	1.10
Heptane	1.14
Di-isobutyl ketone	1.25
Trichloroethane	1.40
Toluene	1.95
Sun Spirits[2]	2.03
Xylene	2.34
"Varsol"[3]	2.34
Kerosene	2.34
"Solvesso"[4]	2.42
"Perclene"[5]	2.50

[1] Product of Carbide and Carbon Chemicals Company
[2] Product of Sun Oil Company
[3] Product of Standard Oil Company of New Jersey
[4] Product of Humble Oil Company
[5] Product of E. I. du Pont de Nemours & Co., Inc.

Moisture

Spunbonded polyethylene sheets are unaffected by either moisture vapor or liquid water. Their properties are the same wet or dry. They are also dimensionally stable to variations in relative humidity, with essentially no change from 0 to 98% R.H.

Temperature

Spunbonded polyethylene sheets melt at about 132°C and in the

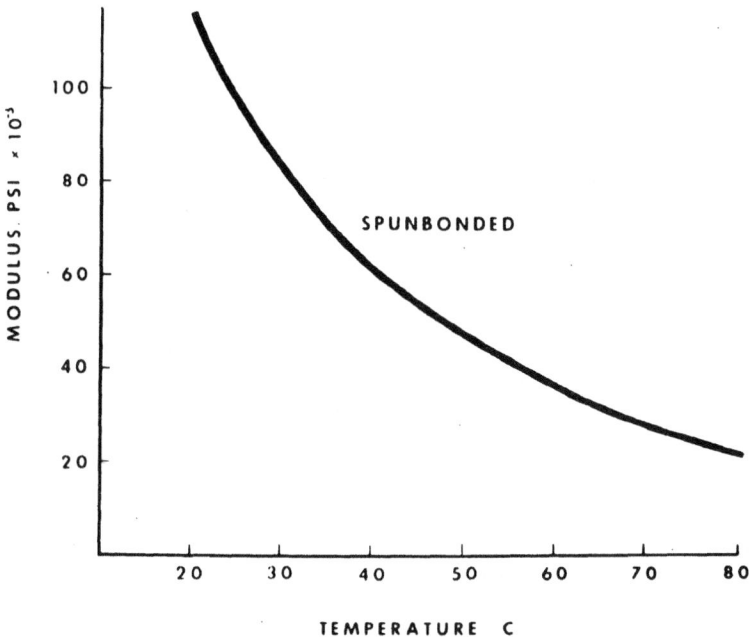

Fig. 3-4. Modulus vs. temperature.

relaxed state begin to lose orientation and shrink at about 120°C. Low temperature resistance is excellent; toughness and flexibility are retained to minus 73°C. Modulus decreases as temperature is increased. Figure 3-4 shows this relationship from room temperature to 80°C. These characteristics are of particular importance where the sheet is exposed to high temperatures in processing, as in laminating, or drying of coatings. Sheet tensions should be kept to the minimum allowed by the equipment to avoid distortion. In general, tensions of about 1.0 lb/in of width are sufficient to maintain control and do not cause sheet distortion in normal elevated temperature processes.

Stability to Ageing and Ultraviolet Light

Accelerated tests indicate good resistance to degradation with age. Physical properties are reduced with extended direct exposure to ultraviolet light from sunlight or fluorescent lamps. Where resistance to such light is important, heavy, pigmented coatings, or light coatings incorporating ultraviolet screeners such as American Cyanamid's "Cyasorb" U.V. 531 can be used to reduce the effect.

Surface Treatment for Adhesion

Spunbonded polyethylene sheets are surface-treated to promote adhesion of coatings, inks, and adhesives by an electrical discharge process similar to that used on polyethylene film. This, together with the fibrous structure, which provides mechanical bonding, results in high adhesion levels for all classes of surface coatings.

Static

Polyethylene is prone to develop static charges when subjected to surface friction. Spunbonded sheets are available with an antistatic treatment to minimize buildup of static charges. When so treated, they can be processed without difficulty in sheet or web handling equipment at most humidities encountered.

Porosity

The air permeability of area-bonded, spunbonded polyethylene is low, in the range of 5 to 17 sec in the Gurley air permeability test, depending on basis weight. Moisture permeability is in the range of 700 to 1000 gm/m^2 (24 hr). The sheets are relatively impermeable to liquid water; a hydrostatic head of 40 to 65 in of water is required to force the first drops of water through a sheet, depending on surface treatment and basis weight. As a result of this combination of permeabilities to water vapor and liquid, a sealed bag containing a deliquescent material such as calcium chloride captures appreciable quantities of water by transpiration of moisture vapor in a humid atmosphere, and holds it as a liquid within the bag.

Conversion Operations—Area-Bonded, Spunbonded Polyethylene

Printing

Spunbonded polyethylene can be printed by all common printing processes, including offset lithography, letterpress, flexography, gravure, and silk screen. Normal inks can be used in all these processes, except where close registry is required, as in four-color litho, or where thick ink films are laid down as in silk screen operations. For these, swelling of the spunboonded polyethylene because of hydrocarbon solvents in the inks causes misregister of the print patterns or, with heavy ink films, puckering of the sheet. Inks, specially formulated with nonswelling solvents to avoid these problems, are commercially available.

Du Pont has also developed a formulation for a water-based ink for Tyvek® adaptable to screen, flexo, and certain types of letterpress printing. This formulation dries rapidly into a tough, adherent ink film. After 4 to 8 days, it becomes water-insoluble. However, it can be readily handled on the printing equipment, and cleaned up without difficulty.

The high whiteness of these sheets results in exceptional clarity and brilliance of the colors laid down in printing. This effect is noticeable even in coated sheets. Most coatings are not completely opaque, and the whiteness of the substrate is evident even through coatings.

Coating

Area-bonded sheets can be coated with a broad range of aqueous and solvent-based coatings on any of the common types of coating equipment, as well as by extrusion. In aqueous systems, acrylics, vinyls, and proteins have been used as binders. Solvent systems based on pyroxylin and polyamide binders are also useful. In extrusion coating, the shrinkage temperature of the spunbonded polyethylene must be taken into account, and only the lower-melting coating materials are applicable, such as branched polyethylene, ethylene-vinyl acetate-wax formulations, and ionomers such as Du Pont's Surlyn.®

Coating formulation is governed by the same considerations that control other substrates, and depends on the end-use requirements of the finished product. These principles have been covered in other publications and are not repeated here. It should be borne in mind, however, that spunbonded polyethylene sheets are not as absorptive as many other substrates and may require somewhat higher binder contents, higher total solids, and higher viscosity than would be used for other materials. In general, the coating of area-bonded sheet is similar to the coating of paper from a process standpoint. Any of the normal coating processes, such as air knife, trailing blade, or reverse roll can be used. Attention should be given to in-process sheet tensions, particularly in operations involving heat, such as drying. For spunbonded polyethylene, sheet tensions should be held to a minimum that will give adequate sheet control to avoid sheet distortion and degradation of "layflat." In general, tensions in the order of 1 lb/in of width are desirable and adequate.

Spunbonded polyethylene, because of the very fine fibers in its structure, holds out coatings effectively. This has a number of consequences in the properties of the coated sheet. These are summarized in Table

3-5, which compares a coated cotton fabric with a coated spunbonded sheet engineered to give equivalent performance. Only about one-third as much coating is required to give a smooth, continuous surface as for the woven cotton. The tensile strength of the spunbonded sheet increases by about 7 lb/in after coating, whereas the elongation and tear strength are essentially unchanged. The woven cotton, on the other hand, shows a substantial loss in both tensile and tear strengths after coating. The coated spunbonded sheet also has superior properties to the coated fabric at less than half the finished weight.

TABLE 3-5
Comparison of Coated Products

| | Spunbonded Polyethylene | | Cotton Sheeting (56×56) | |
| | Area-Bonded | | | |
	Uncoated	Coated	Uncoated	Coated
Basis Wt., oz./yd.^{2}	2.2	—	4.2	—
Coating, oz./yd.^{2}	—	0.8	—	2.3
Tensile, lb/in	46	53	44	35
Elongation, %	28	28	6	9
Tear, Elmendorf, lb	1.3	1.3	2.7	0.62
Thickness, mils	8.0	8.2	11.7	9.2

Coatings can be engineered to give a high degree of abrasion, scrub and stain resistance in the coated product, taking advantage of the toughness and moisture resistance of the spunbonded substrate. High moisture vapor and oxygen barriers can be achieved by coatings of vinylidene chloride. In this case, a sealing coat, such as one of the acrylics, is applied first to the spunbonded sheet. Over this, two coats of vinylidene chloride are applied. With 0.2 oz/yd^2 for each pass, such a coating has yielded moisture vapor transmission rates below 0.2 g/(100 in^2 (24 hr) and oxygen transmission rates below 0.5 cc/(100 in^2) (24 hr) (atm).

Joining and Lamination

Spunbonded polyethylene sheets are compatible with a wide range of adhesives. Good results are obtained with the common water-based adhesives such as the vinyl acetates, animal glues, and starches. As the spunbonded sheet does not absorb much liquid water, it is frequently desirable to reduce the water content of the glue somewhat to speed up development of initial tack and drying. Hot melt adhesives can also be used. Here again, the melting point of the adhesive must be low to avoid heat shrinkage and distortion of the spunbonded sheet. Ethylene-vinyl acetate-wax formulations and branched polyethylene have proved useful.

Heat Sealing

Because the spunbonded polyethylene sheets shrink and de-orient at temperatures below the melting point of polyethylene, it is desirable to precoat the sheets with polymers that soften and join at lower temperatures to achieve good, strong seals. The commonly used bar type or the impulse type sealers provide seals with peel strength as high as 12 lb/in, when the spunbonded polyethylene sheet has been extrusion-coated with either a branched polyethylene or an ethylene-vinyl acetate formulation.

Where desirable or necessary, heat-sealing uncoated sheets can most readily be done with commercially available wedge-type sealers. In this equipment, the two sheets to be joined are clamped between cold jaws with a narrow edge of sheet protruding beyond the jaws. A heated bar is then brought against the protruding edges, melting them together and forming a small bead. The cold jaws prevent shrinkage of the remaining sheet, preserving its initial strength. Such seals attain a strength of 30 lb/in with a 2.7 oz/yd^2 spunbonded sheet.

Vinyl film laminated to a reinforcing spunbonded polyethylene sheet can be dielectrically sealed in the tear-seal type sealers used for unreinforced vinyl film, a process that cannot be used for cotton fabric-backed vinyls.

Embossing

Area-bonded sheet can be readily embossed either cold or at moderately elevated temperatures, around 65°C. Since the sheet, unlike paper, is completely water-insensitive, the embossing patterns are not lost under most conditions of use. With high pressures and a hard, smooth back-up roll, the sheet becomes transparent in the high-pressure areas, leading to a variety of attractive, decorative effects.

Sewing

Area-bonded sheet can readily be sewed in commercial sewing equipment. However, not more than 4 to 5 stitches/inch should be used where seams are subject to high stress, to prevent tearing along the needle perforations.

Cutting

Area-bonded sheets are readily slit, sheeted, and squared on conventional paper equipment. Knives must be kept sharp, because of the toughness and fine fiber structure of the sheet, to prevent rough or distorted edges. Spunbonded sheet can also be die-cut, again observing the need for maintaining the sharpness of the dies.

Typical Uses—Area-bonded Spunbonded Polyethylene

Area-bonded sheets combine in an unique way the stiffness, opacity, and smooth surface of paper, with water insensitivity, and the toughness and strength characteristics of fabrics. With these physical characteristics, they can be converted on paperhandling equipment at high speeds and low cost. This has led to the use of spunbonded sheet in end uses where, in the past, only fabrics have had the necessary strength, but where the use of paper conversion equipment has permitted manufacturing economies.

Uncoated Sheet

Uncoated spunbonded sheet has found application in a variety of end uses such as tags, labels, banners, and signs. Among the tag and label uses, such things as law labels for furniture and bedding, tags for outdoor uses, and pressure-sensitive identification and instruction labels reflect the high toughness and good printability of the product. Outdoor and indoor display and advertising banners and signs take advantage of the toughness and water insensitivity. These are printed either by silk-screen or by offset lithography. Area-bonded sheets have also been used for instruction and specification manuals, where the sheet resists protracted rough handling as a result of its toughness.

Coated Sheet

Coating, in general, adds further abrasion or scuff resistance to the tough spunbonded substrate. Bookcoverings, especially for school texts, and wallcoverings are typical applications. Bookcoverings meet the specifications of the Book Manufacturers' Institute, used as a base by most state education authorities. Wallcoverings based on spunbonded polyethylene have the scrubbability, stain resistance, and ease of removal characteristic of the best vinyl-coated fabric wallcoverings. Combined with this, they also have the styling characteristics of the best paper-based products, and can be converted in mills designed to convert papers. Coated sheets, sometimes combined with aluminum foil as a laminate, are used in heavy-duty packaging. One product of this sort is designed to meet the Mil B-131E specifications, and is used in the packaging of a broad variety of military equipment as well as expensive civilian instruments where a high level of protection is required. Another application is as backing for heavy-duty wet abrasive papers.

POINT-BONDED SPUNBONDED POLYETHYLENE

When spunbonded polyethylene is bonded at discrete points, the resulting sheet is softer, and more drapeable, as was noted earlier. This increased softness results from the greater flexibility of the unbonded fiber network between the bond points. The bonding patterns can be varied to control the final sheet characteristics, ranging over patterns of round points, broken ribs, or fabric-like networks. The bonded area typically is from 2 to 6% of the total sheet area. It is also possible to perforate the sheet completely at the bond points where increased air permeability, for example, is needed.

Physical Characteristics of Point-bonded Sheet

Table 3-6 summarizes the average physical properties of the various

TABLE 3-6
Average Physical Properties
Point-Bonded Spunbonded Polyethylene

Style No.	1421	1443	1458	1461*	Test Reference
Basis Wt., oz/yd^2	1.0	1.3	1.5	1.5	T-410-OS-61
Thickness, mils	5.0	6.0	6.8	7.0	T-441-M-44
Strip Tensile, lb/in MD/XD	10/10	14/15	16/16	11/13	T-404-M-50
Strip Elongation, % MD/XD	27/27	25/28	17/20	40/30	T-404-M-50
Tongue Tear, lb MD/XD	3.0/3.0	3.6/3.1	4.1/4.1	4.4/4.4	ASTM D-2261-64-T
Mullen burst, lb/in^2	47	70	85	83	T-403-M-53
MIT Flex, cycles	>100M	>100M	>100M	>100M	T-424
Opacity, B&L, %	88	89	92	92	T-425-M-60
Frazier Porosity $ft^3/(ft^2)(min)$	<1	<1	<1	20*	Fed. Spec. CCC-T-191B
Handle-O-Meter, grams	23/17	39/18	55/45	35/25	Thwing Albert Company
Delamination, lb/in	0.18	0.20	0.15	—	ASTM D-825-54

* This style is perforated.

styles of point-bonded sheets being offered. Compared to the area-bonded products, these styles are softer and more flexible, higher in tear strength and lower in tensile strength. Point-bonded sheets are also more opaque at a given basis weight than area-bonded sheets. Table 3-7 compares the properties of a representative point-bonded sheet with

TABLE 3-7

Physical Properties of Point-Bonded Spunbonded Sheet

	Spunbonded Polyethylene Point-Bonded	Cotton Sheeting (62×62)	Nonwoven Rayon
Basis Wt., oz/yd^2	1.3	2.7	1.4
Thickness, mils	6	10.2	6.3
Strip Tensile Str., lb/in	14	32.7	4.1
Break Elongation, %	26	12.0	15.0
Work-to-Break in lb/in^2	1.8	1.7	0.4
Init. Mod. x 10^{-3}, lb/in^2	22.4	6.5	29.7
Tongue Tear, lb.	3.3	2.2	0.8
Mullen burst, lb/in^2	70.0	83.4	24.3
Bending length, cm	3.6	2.2	4.4
Opacity, B&L, %	0.94	0.63	0.44
CSIA Abrasion, cycles	44	18	6

All Properties Average of MD and XD

those of a typical woven cotton sheeting and a rayon nonwoven. Although the weight of the cotton sheeting is twice the weight of the spunbonded sheet, the strength properties are similar and the opacity and abrasion resistance of the spunbonded sheet are appreciably higher. In stiffness, point-bonded sheets are intermediate between woven fabrics and paper. Table 3-8 gives a bending length comparison with a woven

TABLE 3-8

Bending Length Comparisons

	Woven Cotton	Spunbonded Point-Bonded	Polyethylene Area-Bonded	Kraft Paper
Basis Wt., oz/yd^2	2.7	2.0	2.2	2.4
Bending Length, cm.	2.2	4.5	6.0	7.2

cotton sheeting, kraft paper, and an area-bonded spunbonded sheet, all at approximately the same weight per unit area. "Bending length" is a fabric-stiffness test that measures the unsupported length to give standardized deflection for a specified strip of fabric. The shorter the bending length, the more flexible the fabric. Flexibility of the spunbonded sheet is sufficiently fabric-like to permit its use in many industrial fabric areas and in certain types of limited-use garments, such as industrial uniforms.

Air Permeability

Air permeability of point-bonded sheet is less than that of area-bonded sheet. The Frazier air porosity of a typical 1.3 oz/yd^2 style is less than 1 ft^3/(min) (ft^2). In a perforated style, the air permeability rises to about 20 ft^3/(min) (ft^2).

Flammability

Point-bonded sheet meets the requirements of the Federal Flammable Fabrics Act for Wearing Apparel (45° angle test). They burn slowly and drip molten polymer. Where flame-proofness is required, the sheet can be coated with flame retardants. When properly chosen and applied in a sufficient weight, burning characteristics approximating flame-proofed cellulosics are possible.

Linting

Laboratory tests indicate that point-bonded sheets do not generate free lint particles under conditions of ordinary use.

General

In most other characteristics, point-bonded sheets are similar to area-bonded sheets. The comments regarding solvents, temperature, ageing, and surface treatments, in the section on area-bonded sheets apply also to the point-bonded varieties.

Conversion Operations—Point-bonded Spunbonded Polyethylene

In general, the comments on the conversion of area-bonded spunbonded polyethylene apply equally well to the point-bonded varieties, with the necessary allowances for the lower stiffness and more fabric-like character of the latter. The following comments in certain areas reflect the more important differences.

Cutting

Point-bonded sheet can be cut by normal fabric-cutting procedures, with reciprocating-knife cutters, in stacks of up to 350 layers. Use of wavy-edged blades and a speed of 1800 strokes/minute is preferable. Straight blades and a 3600 strokes/min speed may result in some fusion between layers at the cut edges, especially in making small-radius cuts.

Multiple-layer die cutting is also possible. Dies, however, must be sharp, again to avoid fusion of the cut edges.

Joining

Spunbonded polyethylene is readily sewed at normal fabric-sewing speeds. Sewn seam strengths are normally about 90% of the base sheet strength. In general, spunbonded sheets can be handled much like woven fabrics.

As with the area-bonded sheet, point-bonded sheet can be joined with adhesives or by heat-sealing. See comments in the previous section for details.

Coloration and Decoration

True dyeing of spunbonded polyethylene is not possible at present; however, the sheet can be colored to medium-dark shades with alcohol solutions of leather dyes, which readily penetrate the sheet structure. Crock fastness and abrasion resistance of these colors can be improved by the addition of about 5% of pyroxylin to the dye solution. Water-based colorant solutions, which give similar results, are also available commercially.

Printing

Printing is a more desirable way of adding color or pattern to the point-bonded sheets. They are readily printed by the flexographic, gravure, screen, or offset lithographic processes. Commercial inks with good adhesion and resistance to crock and abrasion are available. In general, equipment and processes adapted to handling films, coated fabrics, or papers also handle pointbonded sheets.

Typical End Uses—Point-bonded Sheet

The combination of softness, high opacity and high strength available in point-bonded sheet qualifies it for use in many industrial fabric areas. It is of interest in reinforcement and backing applications in furniture, bedding, and seating. It is also used in a variety of disposable industrial protective garments, particularly those where fabric garments become unusable through contamination with paints and adhesives. A related consumer use has been in low-cost swim trunks sold in motels for travellers without bathing suits. Point-bonded sheet is also used in display advertising banners, where a degree of softness and flexibility is desired.

REEMAY® SPUNBONDED POLYESTER

Sheet Structure

Spunbonded polyester sheet products are made up of a random array of continuous, molecularly oriented fibers similar in character to normal textile fibers. The fibers are trilobal in cross section and are in the range of 19 to 25 microns in diameter. Two types of sheet are available: in the first, the fibers are straight, and in the second, they are crimped. Figure 3-5 shows photomicrographs illustrating the difference between these two structures. Because straight fibers have less freedom of movement in the structure, the straight-fiber sheets have a stiff, crisp hand, whereas those with crimped fibers are softer and more conformable. In polyester spunbonded, the fibers are held together by a binder, which can be seen at the fiber cross-over points in the photomicrographs. Generally, the continuous filament structure of the spunbonded sheet permits a relatively low binder content. Figure 3-6 shows that for staple-fiber nonwoven, tensile strength reaches a maximum at 35 to 40% binder, whereas for spunbonded, continuous filament

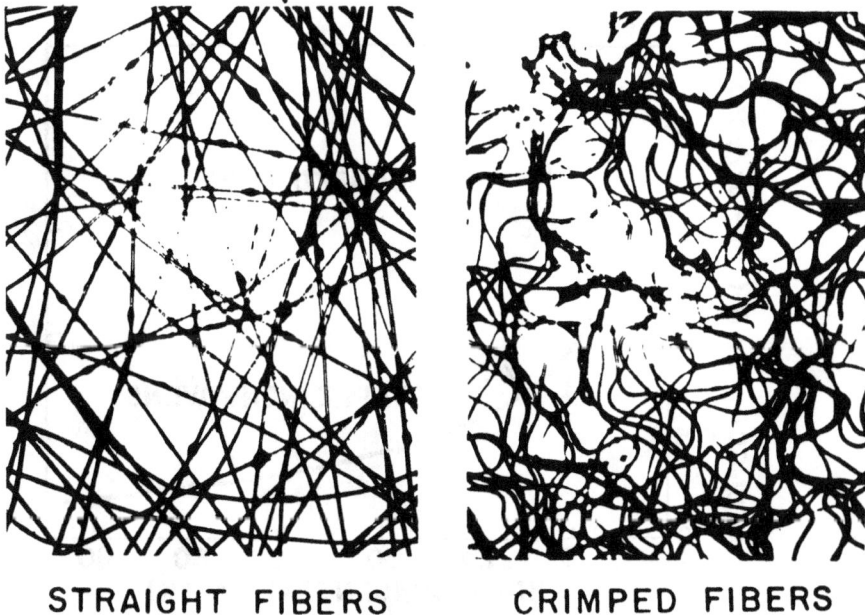

STRAIGHT FIBERS   CRIMPED FIBERS

Fig. 3-5. Differences between structures.

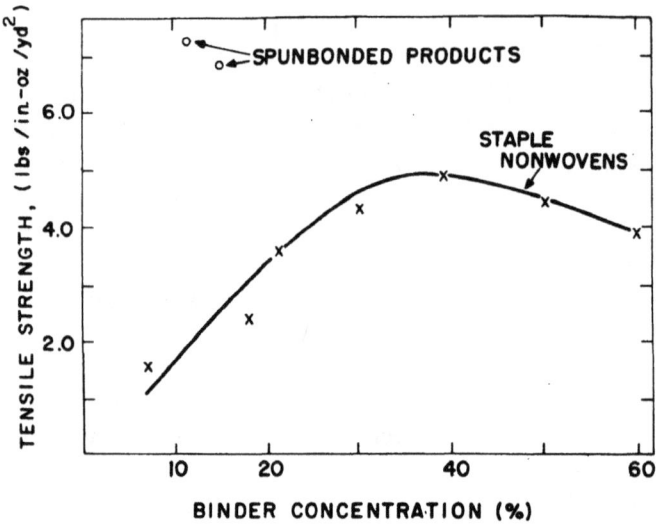

Fig. 3-6. Strip tensile strength.

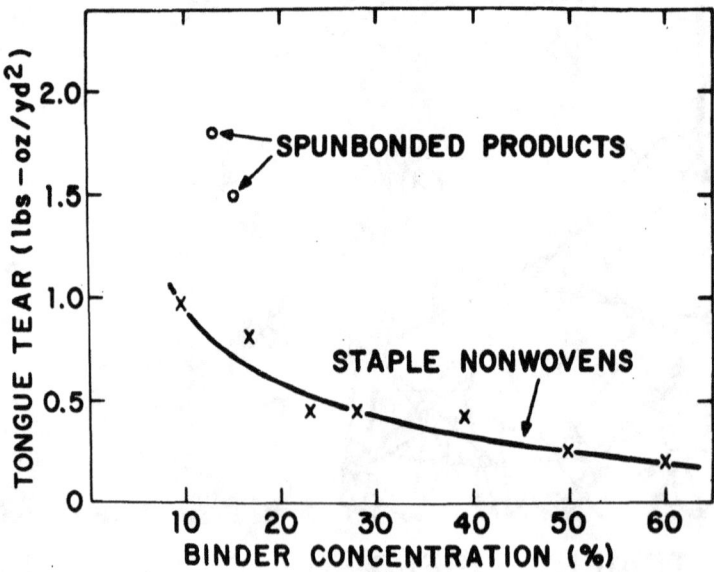

Fig. 3-7. Tongue tear strength.

sheet, even higher tensile strengths are reached with only 12 to 15% binder. Along with the high tensile strength, high tear strength is also achieved at these low binder contents, as illustrated in Figure 3-7. The properties of these straight and crimped fiber types of spunbonded sheet are discussed in greater detail in the following sections.

Physical Properties—Straight-Fiber Sheets

Straight-fiber spunbonded polyester sheets are available in a range of weights from 1.5 to 5.0 oz/yd². The average physical properties over this weight range are shown in Table 3-9. These products are characterized by a combination of high tensile and tear strength, high elongation and high bulk.

TABLE 3-9
Average Properties
Straight-Fiber Spunbonded Polyester Sheets

Style No.	2017	2024	2033	2038	2053
Basis Wt., oz/yd²	1.5	2.2	3.0	3.5	5.0
Thickness, mils	10	12	14	17	23
Strip Tensile lb/in	13/11	25/17	35/25	44/29	66/42
Strip Elongation, %	37/58	72/70	61/67	61/64	71/73
Grab Tensile, lb.	38/29	66/53	106/83	131/95	179/140
Grab Elongation, %	48/56	56/68	69/68	66/70	85/85
Tongue Tear, lb.	2.5/2.9	2.8/3.8	4.5/5.4	6.5/7.9	7.2/9.0
Mullen burst, lb/in²	52	77	115	140	140
Frazier Air Perm., ft.³/(min)(ft)²	340	237	182	191	142

Physical Properties—Crimped-Fiber Sheets

Crimped-fiber spunbonded polyester sheets are available in a variety of finishes and weights ranging from 1.2 to 6.0 oz/yd². Table 3-10 summarizes the average physical properties of these products. In general, crimped-fiber sheets are softer and more drapeable than their straight-fiber counterparts. In physical properties, the crimped-fiber sheets have somewhat higher tear strength and elongation, and somewhat lower tensile strength than the equivalent weights of straight-fiber sheet. Table 3-11 compares the properties of typical straight and crimped-fiber sheets with those of a woven polyester fabric of approximately equal weight.

TABLE 3-10
Average Properties
Crimped-Fiber Spunbonded Polyester Sheets

Style No.	2222	2254	2407	2408	2410	2415	2420	2430	2440	2465	2470
Basis Weight, oz/yd²	2.2	3.5	1.2	1.2	1.3	1.6	1.9	2.5	3.0	5.0	6.0
Thickness, mils	17	21	10	10	10	12	13	17	18	28	30
Strip Tensile, lb/in	15/11	29/20	5.9/3.8	6.2/4.5	7/5	8/6	12/8	17/12	23/16	40/29	55/42
Strip Elongation, %	92/98	83/89	50/53	72/73	60/61	52/58	54/60	72/78	74/74	95/98	94/120
Grab Tensile, lb	45/39	85/68	20/15	20/17	23/17	27/23	42/32	57/42	78/55	116/104	147/126
Grab Elongation, %	87/104	78/92	62/70	69/77	67/72	64/80	63/71	66/84	67/75	89/101	83/109
Tongue Tear, lb	5.0/5.8	6.8/7.6	3.3/3.3	2.8/2.9	3.8/3.9	5.2/5.3	5.8/6.2	6.3/6.5	5.6/7.0	9.2/11.9	11.0/12.9
Mullen Burst, lb/in²	46	82	25	25	26	33	34	64	82	100	134
Frazier Air Perm. ft³/(ft²)(min)	421	248	740	813	780	578	444	309	266	194	120

Table 3-11

Comparative Physical Properties

	Spunbonded Polyester		Woven Polyester (97×83)
	Straight Fiber	Crimped Fiber	
Basis Wt., oz./yd.^{2}	3.5	3.5	3.6
Thickness, mils	19	21	12
Grab Tensile, lb.	113	76	136
Tongue Tear, lb.	7.2	7.2	3.9
Air Permeability ft.^{3}/(min)(ft)^2	191	248	72

Other Physical Characteristics

The characteristics discussed in the following paragraphs apply to both straight-fiber and crimped-fiber types. Where differences exist, they are brought out in the pertinent paragraphs.

Solvent Resistance

Spunbonded polyester sheets have excellent resistance to organic solvents such as hydrocarbons, alcohols, and ketones, together with good resistance to halogenated solvents such as trichloroethylene and perchloroethylene. They also have good resistance to acids and alkalies over the range from pH 0.1 to pH 10.0 at room temperature.

Moisture Resistance

Moisture pickup of spunbonded polyester is only 0.5% at 98% relative humidity. The sheets are highly dimensionally stable over the entire range of humidity. Physical properties are essentially unchanged—wet or dry. Spunbonded polyester also exhibits low shrinkage under typical home laundry conditions, together with good stiffness retention after washing.

In common with most synthetics, these products are highly resistant to degradation by molds, mildew, rot, and perspiration.

Temperature

Spunbonded polyester sheets melt at approximately 250°C. They can be processed readily at temperatures below 177°C, but temperatures above 210°C should be avoided. As these sheets are thermoplastic and have high break elongations, they can be molded to make formed articles.

Stability to Ultraviolet Light

These sheets are similar to other polyester fiber products in their resistance to sunlight. In tests in Florida sunlight, spunbonded polyester sheets retained approximately 80% of their initial tensile and tear strengths when exposed for six months under glass, and about 20% when exposed directly.

Static

Spunbonded polyester sheet accumulates an electrostatic charge when passed over rolls, spreader bars, or other control surfaces. These charges can build up sufficient potential to cause spark discharges, which must be controlled, particularly when flammable vapors are present. Control measures are well defined, and include thorough grounding of all process equipment, rolls, ducts, and other machine parts, and the use of any of several commercially available types of static eliminators at critical points. Where practical, one of the most positive means of preventing build-up of electrostatic charges is to maintain the relative humidity above 60% in processing areas through the use of steam or water sprays.

Ravelling

Cut edges of spunbonded polyester sheets do not unravel or fray. Binding or sealing of cut edges is not necessary.

Softness and Drape

Bending length is a commonly-used test for measuring the relative softness of sheet products and fabrics. This test measures the free length of a strip of material needed to reach a standard deflection. The lower the bending length, the softer the structure. Figure 3-8 shows the spread in bending length for the various types of spunbonded polyester plotted against sheet weight. The crimped-fiber products fall toward the lower part of this spread; the straight-fiber products tend toward the upper. Curves are also shown for typical staple non-woven sheets and for typical woven cotton fabrics.

Conversion Operations—Spunbonded Polyester Sheets

The following comments on conversion operations apply to both straight and crimped-fiber products. Process adjustments are in general required only to accomodate the varying sheet stiffnesses.

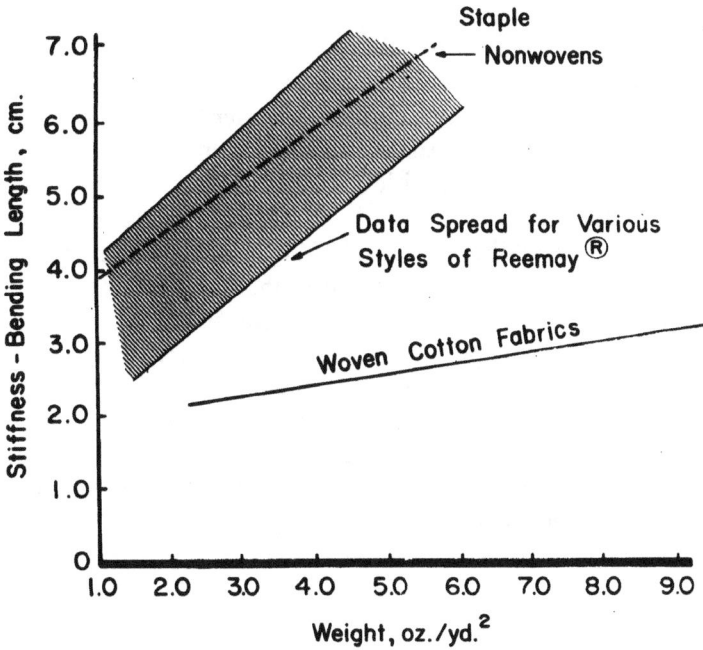

Fig. 3-8. Stiffness vs. weight.

Textile Operations

Spunbonded polyester sheet can be successfully treated in a variety of standard textile processing operations. These include dyeing, printing, sanding, sueding, embossing, calendering, slitting, die cutting, and sewing.

Printing and Flameproofing

Spunbonded polyester sheet is readily printed with standard fabric roller-printing equipment. Tensions through the press should be held to the minimum operable level, particularly with the lighter weight sheets, to avoid necking-down and wrinkling. Excessive pressure or set on the print rollers should also be avoided because these cause widening of the sheets and result in poor registry of the pattern. These products can also be printed on web-fed gravure and flexographic presses such as are used for printing films and papers, with similar precautions as to tensions and print-roll pressures.

Frequently flameproofness is required in the final, printed product, particularly in disposable end uses. To achieve flameproofness, the printing pastes are modified by using flame-retardant binders and the printed sheet is topcoated with additional flame retardant. Printing pastes are normally formulated from a pigment for coloration, a binder for adhesion and crock resistance, and an extender for viscosity control. For flame-retardant printing pastes, polyvinylidene chloride or polyvinyl chloride binders, with addition of 10 to 15% tris 2,3-dibromopropyl phosphate, are recommended. Following printing, the sheet should be coated with additional 15 to 20% by weight of flame retardant by padding to achieve durable flame resistance and crock and washfastness. A typical top coating contains about 50% tris 2,3-dibromopropyl phosphate and 3% polyvinyl chloride in water.

In some applications, it is desirable to have a smoother, less fibrous printing surface than the spunbonded polyester sheet presents. In these cases, the sheet can be filled or coated with a wide variety of coatings such as are described in the next section. Such filled or coated sheet can be printed by any of the usual printing processes. The printing characteristics depend primarily on the coating used rather than on the characteristics of the base spunbonded sheets.

Coating

Spunbonded polyester sheet can be coated with a wide variety of coatings and by essentially any commercial coating process.

Coatings can be applied from aqueous systems with acrylic, vinyl, protein, or acetylated starch binders. Solvent coating systems with pyroxylin and vinyl chloride binders have also given similarly good results. The high melting point of the polyester spunbonded also permits the use of plastisol systems based on vinyl chloride, either compact or expanded, and of extrusion coating with common extrusion resins.

As with all thermoplastic materials, tensions should be held to the minimum operable levels, where the sheet is heated, to avoid stretching, necking, or distortion. Experience has shown that unsupported sheet can be processed without stretching at sheet temperatures up to 149°C. With support, the useful maximum sheet temperatures can be raised to about 204°C. Equipment temperatures can exceed these maxima considerably where sheet is being handled at high speeds and sheet temperatures do not reach the equipment temperatures.

Joining and Laminating

Adhesives Spunbonded polyester sheets can be readily joined or laminated to other materials with a wide variety of adhesives. Where the maximum adhesion is required, back-coating with natural or synthetic rubbers, or with acrylics may be desirable. Such back-coating formulations are commercially available from a number of suppliers. In other applications, where "quick tack" is desired with latex adhesives, the spunbonded polyester sheets can be coated with aluminum sulfate prior to joining to promote this characteristic. Aluminum sulfate should be applied at the level of 4 to 8 % based on the weight of the spunbonded sheet. It may be applied from water by a variety of processes, including spraying, padding, or knife coating, with the addition of a film-forming thickener such as Hercules' "Natrosol" 250 HHR hydroxyethyl cellulose.

Sewing Spunbonded polyester sheets are readily sewn by conventional techniques. Cut edges do not ravel or fray, which can in some instances simplify seaming through elimination of binding or overcasting of edges.

Cutting

Spunbonded polyester sheets can be cut by all conventional techniques, including slitting, sheeting, guillotine, knife, or die cutting. In any cutting process, cutting edges should be kept sharp and free of nicks or other defects.

Coloration

Spunbonded polyester sheets can be dyed by the procedures and with the dyestuffs normally used for polyester fabrics.

End Uses

Spunbonded polyester sheet is finding application in a variety of uses, many of them industrial in nature. It is used as interlinings in garments such as women's blouses and dresses, men's shirts, pajamas, robes, and cloth hats. These uses reflect the bulk at light weights, and the retention of stiffness and dimensional stability after repeated washings. In the shoe industry, these products are used as backings for leather and vinyl shoe linings, and for reinforcing components. These uses take advantages of the rot- and mildew-resistance, nonfraying and moldability of the spunbonded sheet. Polyester fibers have attractive electrical characteristics, and thin, calendered, spunbonded sheet is used in electrical insulation in motors and cables. Spunbonded sheet is an attractive reinforcement for vinyl films, either applied as a laminate or formed by coating or extrusion.

TYPAR® SPUNBONDED POLYPROPYLENE

Sheet Structure

Typar® spunbonded polypropylene sheet products are similar to the spunbonded polyesters in that they are made up of an array of continuous, molecularly oriented fibers similar in characteristics to normal spun textile fibers. The fibers are circular in cross-section and are about 40 microns in diameter. Fibers in the random web are straight, as shown in the photomicrograph of Figure 3-9. Spunbonded polypropylene sheets are self-bonded at the filament crossover points.

Physical Characteristics

Spunbonded polypropylene sheets were developed for use as the primary backing for the manufacture of tufted rugs and carpets, which has become the major carpet-making process in this country since 1953. A further description of this process will be found in a later section of this chapter. The physical properties of spunbonded polypropylene sheets have been engineered to meet the needs of the carpet

Fig. 3-9. Structure of spunbonded polypropylene.

industry.   They are available in weights ranging from 2.5 to 3.5 oz/yd^2.
Table 3-12 summarizes the average physical properties over this weight
range.   These products are characterized by high tear strength and
toughness.   Table 3-13 compares a typical spunbonded polypropylene
with a woven jute fabric.

TABLE 3-12
Average Physical Properties
Spunbonded Polypropylene Sheets

Style No.	3012	3300	3350	Test Method
Basis Wt., oz/yd^2	2.5	3.0	3.5	T–410–OS–61
Thickness, mils	9	11	13	T–441–M–44
Strip Tensile, lb/in	30/23	26/33	32/32	T–404–M–50
Strip Elongation, % MD/XD	42/24	39/37	37/29	T–404–M–50
Grab Tensile, lb	82/88	117/130	131/157	ASTM D–1682–64
Grab Elongation, %	49/43	75/65	70/67	ASTM D–1682–64
Tongue Tear, lb	8/10	14/15	16/18	ASTM D–2261–64T
Elmendorf Tear, lb	7/5	10/13	11/13	T–414–M–49
Mullen Burst, lb/in^2	167	183	222	T–403–M–53
Spencer Puncture, in-lb/in^2	66	85	101	Thwing Albert Manual
Opacity, B&L, %	43	45	59	T–425–M–60
Air Permeability, ft.^{3}/(ft.^{2})(min)	111	78	68	Fed. Spec. CCC–T–191–B

TABLE 3-13
Comparative Properties
Spunbonded Polypropylene And Woven Jute

	Spunbonded Polypropylene	Woven Jute (15 × 10)
Basis Weight, oz/yd^2	3.5	10
Grab Tensile, lb	143	106
Tongue Tear, lb	17	53

Moisture Resistance

Spunbonded polypropylene is insensitive to moisture, retains its
physical properties, does not shrink when wet, and is resistant to rot
and mildew.   It does not stain other yarns in contact with it, as in
carpet construction.

Temperature

Spunbonded polypropylene melts at about 170°C. It has good dimensional stability below 132°C, with less than 1% shrinkage on heating to that point. Drastic shrinkage takes place at temperatures above 150°C.

Conversion Operations

As was pointed out earlier in this section, spunbonded polypropylene was developed as a primary backing for tufted carpets. Its conversion is therefore described in this context. The great majority of all carpets are now manufactured by tufting, rather than by weaving. The basic elements of the tufting process are illustrated in Figure 3-10. The yarn that is to form the pile of the carpet is threaded through a large needle which is pushed through the primary backing, where a looper catches the yarn and holds it as the needle is withdrawn, forming a loop on the top face of the carpet. The backing is then advanced, and the process repeated to form a closely spaced row of yarn loops. In the

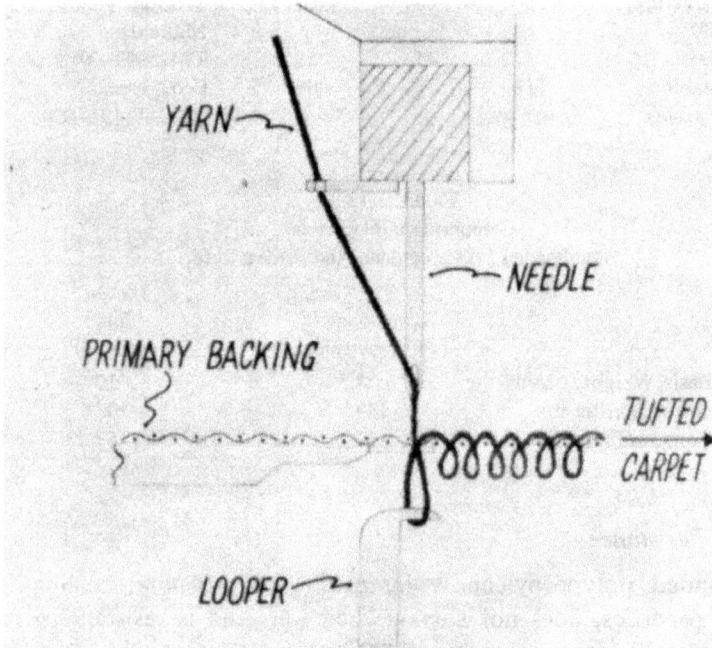

Fig. 3-10. Tufting process.

tufting machine, there may be as many as 1800 such needles and loopers in a row, working in unison to form the pile in a 15-ft wide carpet.

Following tufting, another, usually woven, fabric is glued to the underside of the carpet with a latex adhesive to lock the tufts in place and provide additional stiffness in the finished carpet. The second fabric is called the secondary backing. The primary backing is the structural base of the carpet. It holds the tufts in place and provides the strength and dimensional stability in the finished carpet. Until recently, these primary carpet backings were solely woven jute fabrics, in the weight range of 8 to 12 oz/yd². Spunbonded polypropylene sheets, which provide the same functional properties in the weight range from 2.5 to 3.5 oz/yd², have been developed. These sheets also have a number of additional advantages such as rapid drying, freedom from mildew, rot, and insect attack, dimensional stability, freedom from staining of the pile yarns, more uniform formation, and closer possible spacing of the yarn loops.

Many of the advantages of spunbonded carpet backing result from the arrangement of continuous fibers in the structure. In a woven

Fig. 3-11. Fiber "collar".

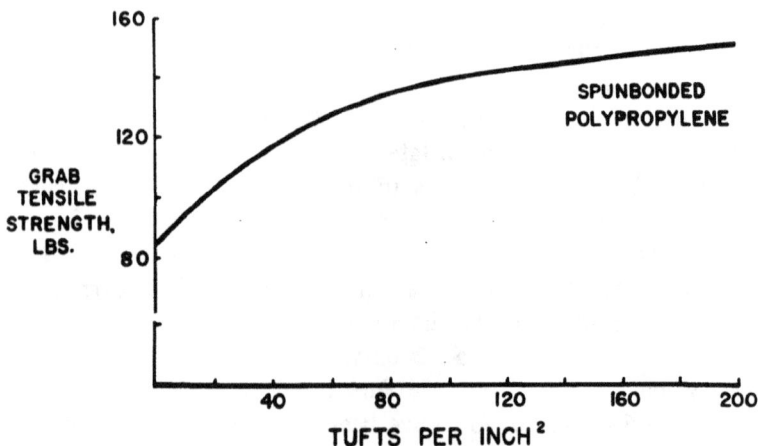

Fig. 3-12.   Strength vs. tuft density.

backing, when the needles penetrate the structure, they may enter either between the yarns or through them.   In this process, either the needles may be deflected or the backing yarns may be partially broken.   Needle deflection leads to nonuniform placement of the loops and distortion of the loop pattern.   Breakage of the backing yarns leads to reduction in strength of backings.   With spunbonded backings, in contrast, there are no discrete yarn fiber bundles to deflect the needles and the pattern of loop placement is more uniform.

Further, the penetration of the needles deflects the individual fibers of the structure so that a "collar" of fibers is formed around each tuft, without any fiber breakage.   This effect is shown in the photomicrograph in Figure 3-11.   These fiber "collars" have a reinforcing effect, and spunbonded sheet is stronger after tufting than before.   The increase in strength becomes greater as the density of tufts increases, up to about 180 tufts/in^2.   This effect is illustrated in Figure 3-12.   This phenomenon permits carpets to be made with a greater density than is feasible with woven backings, a desirable characteristic where the maximum wear resistance is desired in the carpet.

SUMMARY

Spunbonded sheets offer a wide range of product characteristics, ranging from very light weight, flexible structure, to heavy, stiff materials.   All characteristically, have high strength-to-weight ratios compared to other nonwoven and woven structures.   They range from sheets with all the appearance characteristics of papers, with high opacity per unit weight, to open, fibrous products.   Being wholly synthetic, all share insensitivity to water and resistance to the degrading effects of rot, mildew, and insect attack. All are readily converted in a wide range of standard commercial processes and equipment. They represent a new class of man-made products, falling between papers and films on the one hand, and the woven fabrics on the other. In a real sense, they offer a range of new property combinations, previously unavailable to converters, to make economical products meeting a wide variety of end-use needs. As they find their place in the market, they will undoubtedly undergo changes, and will be offered in new variants to meet the needs of industry better.

chapter 4

Nonwovens and Textile Replacements

ALBERT D. GUSMAN

INTRODUCTION

This chapter is designed to provide an ever-increasing number of interested persons with broad and general understanding of nonwovens and their various interlocking aspects. It would not be feasible to provide the reader who lacks technical background with insights into all the basic methods of manufacturing a nonwoven or a textile replacement. Therefore this chapter proposes to cover in a broad way the general nonwoven industry as it is now emerging.

To discuss in detail here the specifics of nonwoven processing would mean condensing a tremendous amount of material covering the various processes and applicable machinery and the peculiarities of the various families in the chemical, fiber, paper and textile industries. In addition, there is an extensive overlapping of the technology that is employed in each of the above industries that provide a group of "new" nonwovens or textile replacements. Paper, textile, fiber and chemical companies see a marriage of many of the materials or substrates produced in a variety of combinations selling to many markets in competition with each other!

Economic forces today and demands for new sophisticated nonwoven materials in various industries, challenge the imagination and engineering skills of machinery builders, fiber companies, paper, textile, and chemical firms, to produce new machines, new systems or new processes. All these firms can assuredly look forward to developing new profitable equipment and products for this new growth industry.

BACKGROUND

Paper making is 3,000 years old. Since ancient days, paper has been made by having natural vegetable fibers interwind in water in thin leaf form. With the development of modern industrial technology, paper began to be manufactured by using chemical and synthetic fibers as principal and auxiliary raw materials and using a binder made of synthetic resin or the like to replace the old binding method of interwinding in water the fibers formed into fibrils by beating. Thus, an advanced "Synthetic" paper has come to be manufactured.

"Nonwoven fabrics" is a new department of the fiber industry rapidly growing in the last several years, after 4,000 years of textiles made by a combination of spinning-weaving techniques with chemical processes. Nonwoven fabric is manufactured by using natural, chemical and synthetic fibers as raw materials, making webs by means of air current or garnett machines under a dry process, and then forming a layer of cloth from the webs by use of a binder.

The synthetic fiber paper and the nonwoven fabric, briefly described above, resemble each other in appearance of product, but they are manufactured in different branches of industry, the former in the paper industry and the latter in the fiber processing industry. The synthetic fiber paper is usually manufactured by means of a wet process, forming the webs in water. In the manufacture of nonwoven fabric, the webs are generally formed through use of a dry process.

In some quarters, there is a tendency to classify the synthetic fiber paper as a nonwoven fabric manufactured by means of the wet process. As for the characteristic features of the wet and dry process, the former permits a high manufacturing speed and low cost, whereas the latter process makes a product similar to cloth by using longer fibers.

The first chemical-synthetic fiber paper appeared in 1928 when the manufacture of chemical fiber paper through the utilization of viscose rayon was patented in Germany. In 1950, paper of 100% glass fiber was successfully manufactured. In 1953, paper composed of 100% Dacron was produced, and in 1954 synthetic fiber papers consisting of 100% Nylon, Orlon, and Dynel were manufactured.

TYPES OF NONWOVENS

Nonwovens are produced on a variety of textile machines, modified paper machines, needle machines, and by many combinations of these processes. Of recent commercial significance in the field is the development of spun- or spinbonded nonwovens that are manufactured from polymer to fabric on a continous basis (see chapter 3). Coupled to this, we have a series of new sophisticated systems, now in the development stage, that will be commercialized by 1975.

The significance of these different production methods is that they are competing in many instances for the same markets. The desirability of developing some "unique material or process" has created a "crash" program by many top flight firms in the United States and abroad as witnessed by the listing of patents thus far issued. An examination of these patents further points out the significant trend and potential profitable products these firms anticipate in the future growing markets. Some of the typical nonwovens produced today are briefly described as follows:

(1) *Conventional saturated type* employing Garnetts, Curlators or other textile machinery to produce a variety of lining substrates for such uses as shape retention in fashion apparel. These nonwovens are not limited to fashion. They may employ different bonding agents and combinations of fiber to produce air filters, and can produce shoe materials for applications such as box toes, counters, and backing materials for coated fabrics.

(2) *Conventional high-loft materials* are being produced on Curlators but instead of being saturated and cured in an oven, they are carefully sprayed with a variety of binders to maintain their lofty quality for such uses as high loft air filters, quilting fillers and a host of other products. In addition to this, by the application of different grades of silicon carbide or aluminum oxide, they become very effective scrubbing pads or floor polishing pads and buffing pads. 3M company, Norton, Loren, and others are actively engaged in this very profitable business.

(3) *Scrim reinforced nonwovens* are profitably being produced by some paper firms such as Kimberly-Clark and Scott Paper. Here is an instance where a nonwoven or cross-laid scrim is "laminated" to a variety of different layers of tissue to produce an excellent disposable substrate. This does not imply that other means may not also be employed to produce such disposable materials. For example, Chicopee Mills division of Johnson and Johnson, has for many years produced unidirectional light nonwovens for use in disposable products. Likewise, Kendall Company employed this technique and others do so today very effectively. The convergence of technologies employed to produce a variety of disposable substrates is limited only by the imagination

and creative talents of those firms engaged in this growing business. Johnson & Johnson has a facility to produce tissue substrates that aid them in the development of new products, while at the same time they employ conventional machinery to manufacture nonwovens on textile machinery.

(4) *Wet laid nonwovens* are manufactured on paper machinery that in some instances has been modified to accommodate fibers longer than are normally used in papermaking. In addition, new substrates have been produced with pulp and fiber combinations. Dexter Company, Schweitzer division of Kimberly-Clark Company, Ecusta division of Olin, and others also have this capacity.

(5) *Needled nonwovens* employ a needle loom to "bond" fibers mechanically to produce such products as nonwoven blankets, indoor-out door carpeting, needled nonwoven substrates employed as a primary backing for tufted carpets, and a host of other products.

(6) *Tissue fiber laminates* where fiber is carefully spread to manufacture a composite structure is another technique employed here in the United States and abroad. The characteristics are somewhat different from those of other nonwovens.

(7) *Spunbonded* or spinbonded (as referred to in Europe), essentially produces nonwoven structures from polymer directly to fabric. Unusual characteristics have been developed by means of this technique, so that by variation of the polymer, the end use might be a primary backing for tufted carpets, a disposable substrate, a fashion interliner, or a host of other products.

Of the new emerging technologies, spunbonded substrates offer a unique variety of properties not obtained by the other nonwoven materials.

Spunbonded products are sheet structures composed of randomly arranged filament fibers bonded primarily at filament crossover points.

These sheet structures are made by a unique process in which spinning, web formation, and bonding are highly integrated.

Spunbonded products represent a new class of fibrous materials and offer important performance parameters different from those of conventional textile materials. Their properties can be varied over a wide range so that they can be expected to compete with certain classes of textiles, films, and papers as a direct textile or other material replacement! (See chapter 3.)

"Reemay", "Typar" and "Tyvek"

"Reemay" is based upon a polyester polymer, "Typar" on isotactic polypropylene, and "Tyvek" on high-density polyethylene.

Obviously, their chemical and thermal properties typically reflect the chemical and thermal properties of the respective polymers.

"Reemay" is made from either straight or crimped fibers having filament deniers in the range of 2 to 4. The polyprolylene filaments in "Typar" are straight and about 8 dpf. In "Tyvek" the polyethylene continuous filaments are very fine, ranging from 0.08 to 0.8 mil (less than 1 dpf). In all cases, the stiffness or drape of the sheet is dependent upon the bonding conditions that are selected to meet the specific end use requirements of the product.

Both "Reemay" and "Typar" are stronger than cotton fabrics of comparable weight. Because of its higher filament denier and degree of bonding, "Typar" may be considered stronger, especially in tear strength, than the other two spunbonded products. The aesthetics of "Reemay" and "Tyvek" approach those of woven and knitted fabrics more closely than does "Typar". All spunbonded products exhibit little if any bias stretch because the filaments are randomly oriented within each of the sheet structures. This does not imply that at a later date this quality cannot be altered.

Some end uses of "Reemay."

Apparel interlinings

Shoe components

Tags and Labels

Substrate or Backing for Coatings—The ready acceptability of coatings, the bulk at light weights and the superior embossability, provide styling advantages coupled with excellent strength and durability in bookbindings.

Filter Media—The high porosity coupled with good retention characteristics and low pressure drop are being utilized in both liquid and dry filtration applications.

Electrical Insulation—As a smooth calendered product, "Reemay" offers structures as thin as 3 mils. These are being used in combinations with "Mylar" for electrical motor insulation.

Reinforced Plastics—"Reemay" is readily wetted out by polyester and epoxy resins, which are used in both the electrical and the reinforced plastic industries. In high-pressure laminates as a reinforcement for epoxy resins, improved electrical properties have been demonstrated over those of comparable glass fabric reinforced products.

"Typar"—primary applications today are for use as a primary backing for the tufted carpet industry but are not limited to this one application.

Some end uses of "Tyvek"

Bookcovering	Tags	Charts
Wallcovering	Abrasive backing	Utility and Construction sheeting
Banners and Signs	Packaging	Disposable Apparel
Posters	Maps	

(8) *Stitch-Bonding Nonwovens*

Type A- "Arachne"

This technique was designed to create improved nonwoven fabrics for use in the outerwear field, as well as other products, including blankets, hospital fabrics, and industrial and asbestos fabrics. The principle is that of stitch-bonding a continuous fiber web, and simultaneously knitting through the webs. The web is reinforced by warp knitting in one of the basic "weaves" or in combination of "weaves". The Arachne machine essentially operates like a warp knitting machine.

Type B—Maliwatt Stitch Bonding

This machine uses the sewing principle for the manufacture of novel textile structures thus making use of the high sewing speeds obtainable. The manufacturing process is either a continuous one or it is divided into two distinct phases (the web-forming operation and the stitch-bonding process). If the discontinuous process is used, the web is put up in lap form on the Maliwatt machine which reinforces it by a plurality of parallel warp stitch chains.

More economical production is achieved by the continous operation whereby the webs are formed and cross-lapped and then fed into the stitch part of the machine.

Type C—Malimo Stitch Bonded Materials

This is a new type of machine characterized by stitch-bonding crossed warp and filling systems of threads, thus creating plane textile structures.

Fig. 4-1. 360″ Model 20 Fiber/Locker needle felting machine, with a right-hand open gate for needle-punching endless papermaker's felts. *Courtesy of James Hunter Machine Co., Div. of Crompton & Knowles Corp.*

Fig. 4-2. 105″ Model 24 Fiber/Locker needle felting machine all-purpose unit for needle felting blankets, carpets, heavy industrial felts. *Courtesy of James Hunter Machine Co., Div. of Crompton & Knowles Corp.*

Fig. 4-3. 105″ Model 16 Fiber/Locker needle felting machine for needle-felting nonwoven blankets and light technical felts. *Courtesy of James Hunter Machine Co., Div. of Crompton & Knowles Corp.*

Fig. 4-4. 500″ Model 25 papermaker's felt machine *"world's largest needle felting unit"* installed at Draper Brothers, Inc., Canton Mass. Illustrated with gate open for removal of endless felt. *Courtesy of James Hunter Machine Co., Div. of Crompton & Knowles Corp.*

Fig. 4-5. 310″ batt making and papermaker's felt, Knox Woolen Co.,
Camden, Maine. Illustrated from right center to left:— Tandem
weighing feed units, 2-cylinder Garnett, crosslapping system and
floor apron assembly with pneumatic compressor. Inclined
apron delivering batt to the feed apron of 310″ Model 19
Fiber/Locker.   *Courtesy of James Hunter Machine
Co., Div. of Crompton & Knowles Corp.*

Fig. 4-6. 105-Model 22 Fiber/Locker   needle felting machine for needle
felting nonwoven blankets. *Courtesy of James Hunter Machine Co.,
Div. of Crompton & Knowles Corp.*

Fig. 4-7. Swinging apron of 598 Crosser Lapper, 240″ wide. *Courtesy of Proctor & Schwartz.*

Fig. 4-8. 600 Garnett for cotton & synthetic batting. *Courtesy of Proctor & Schwartz.*

Fig. 4-9. Duo-Form to produce airlay webs. *Courtesy of Proctor & Schwartz.*

Fig. 4-10. Side view of Duo-Form showing airlay section, carding section and feed. *Courtesy of Proctor & Schwartz.*

Fig. 4-11. Delivery end of Duo-Form. *Courtesy of Proctor & Schwartz.*

There are other methods of producing novel nonwoven substrates that are actually combinations of some of the typical types described in this chapter.

Figures 4-1 through 4-11 show machines for producing nonwovens.

Now let us examine some of the significant potentials in the growing field of disposables.

DISPOSABLES

One must carefully examine the use requirements of nonwovens and the manner in which they are manufactured to understand that the function of disposability may be achieved by more than one technology. Every major paper and fiber company is examining the role it will play in this vast growing business.

Coupled to this, key textile and machinery manufacturers are searching how they will best serve this new industry. Nonwoven disposables are a big business right now yet the disposables industry is still in its infancy. Recently it has been recognized as a separate and new entity. The Disposables Association has now been organized and formally announced. Originally, there were only 24 corporate members. Today, the association has 122 corporate menberships, and its open meetings in New York City have drawn over 300 persons representing many industries all with a vested interest in disposable soft goods. Memberships include major corporations in textiles (fibers, mills, converters), paper, chemicals, plastics, machinery manufacturers, and fabricators, and include foreign representation. It is probably the single trade association that brings together so many diverse industries seeking a common goal: propagation of the disposable soft goods.

One disposable item that has been in institutional use for some time is an absorbent bed pad for hospitals. In 1968, approximate estimates indicated that sales of absorbent bed pads to American hospitals approached the $20 millon mark.

There are about 1.8 million hospital beds in this country. Each patient requires fresh bed linen daily; frequently, fresh examination gowns, X-ray capes and other items. Disposable items for these particular uses are already in limited use. It is only a matter of time before they are more broadly accepted by institutions as standard supplies. Economic factors are pointing to disposables to ease laundry costs. (See Table 4-1)

TABLE 4-1
**Summary of Disposable Materials Required
to Penetrate 1% Of Present, Selected
Institutional Markets**

		Square Yards	Pounds
1.	Bed Sheets	195,025,000	32,500,000
2.	Pillowcases	25,190,000	4,198,000
3.	Draw Sheets	10,735,000	1,790,000
4.	Head Rest Covers	163,800	27,300
5.	Terry Bath Mats	1,843,250	921,600
6.	Terry Wash Cloths	2,386,090	596,500
7.	Terry Bath Towels	20,031,200	6,677,000
8.	Terry Hand Towels	4,643,600	1,537,870
9.	Huck Hand Towels	11,563,600	2,890,900
10.	Autoclaved Items	1,981,250	372,000
11.	Medical Exam Gowns	11,497,500	1,916,250

Envision such items as surgeons' gowns, nurses' and anesthetists' gowns, head covers, face masks, gloves, and conductive shoe covers becoming universally standard hospital inventory in the future, replacing durable items. Rayon is the dominate fiber, either in 100 per cent rayon constructions, or in structures with cellulose, where rayon fibers act as reinforcing agents to improve tear and tensilestrengths.

At this early stage, there are not enough disposables in total use in hospitals to permit an accurate survey on comparative costs of disposables versus durable items that require maintenance. When that day comes, it may be possible to prove that by replacing a durable bed sheet, for instance, with a disposable bed sheet one can lower operating costs.

Hospitals face another problem with durable linen—pilferage. The solution a disposable offers is obvious. First, the minimal value of a disposable lessens the inclination to pilfer. Second, should pilferage occur, the loss of a 30-cent disposable sheet cannot be compared with that of a $2.00-muslin sheet.

One look at the potential shows that disposables in the hospital-institutional field offer many new profit opportunities. With regard to the approximate 1.8 million hospital beds, consider how this figure must increase steadily because of availability of Medicare to thousands of new patients. As these increased burdens are forced upon the already overcrowded facilities and overworked staffs of the medical institutions, hospital labor costs increase once again.

The hotel-motel and transportation fields offer significant opportunities for disposable soft goods. Take bed sheets: If only 1 per cent of the total institutional bed sheet. market were penetrated by disposables, over 195 million square yards of nonwoven fabrics would be consumed in one year.

Another hospital disposable item is the complicated operating room pack, which contains about 10 square yards of nonwoven disposable materials. It is estimated that there are about 13 million operational procedures performed in this country annually, with slightly less than 10 per cent penetration of this "market" using disposable packs at the present. That 10 per cent penetration represents consumption of another 13 million square yards of nonwoven disposable materials.

LINEN SUPPLY FIELD

After looking at the potential for disposables in the institutional market place, now consider the industrial community. The linen supply industry originally viewed disposables with a very wary eye because of fear that their rental business would dwindle as disposables grew. However, most operators are smart enough to realize that their most valuable assets are their established lines of distribution: their customers and their method of reaching those customers—their trucks. It is not that they are renting, but that they are servicing industry with needed products. So does it matter, after all, whether they service those industries with woven coveralls or with disposables?

Now look at the size of that rental market for industrial linen, and the potential it offers for disposable replacements. Figures compiled by the Linen Supply Association of America for 1966 show 22 categories of industrial items and their unit rental totals for that year. (See Table 4-2)

TABLE 4-2
Durable Rental Items Supplied to
Industrial Marketplace, 1966

Item	1966 Unit RENTALS
Aprons, Bib/Bar	466,720,000
Coats, Short	42,270,000
Coats, Long	69,230,000
Dresses	85,870,000
Pants	96,111,000
Shirts	112,119,000
Overalls, Coveralls	18,148,000
Tablecloths	204,463,000
Napkins	1,892,088,000
Towels, Barber	1,053,008,000
Towels, Bath	519,672,000
Towels, Kitchen/Bar	1,246,441,000
Towels, Massage	911,995,000
Towels, Office	440,400,000
Towels, Continuous	74,993,000
Hair Cloths	18,566,000
Dust Mats	10,900,000
Dust Mops	23,121,000
Wiping Cloths	364,064,000
Sheets	503,866,000
Pillowcases	318,426,000
Miscellaneous	147,565,000
TOTAL	8,620,036,000

With these figures as a base, American Viscose selected four items (pants, shirts, long lab coats, aprons) and made its own tabulations.

TABLE 4-3
Projected NonWoven Yardage for Disposable
Industrial Garments in Selected Categories

Items	Unit Rentals	Sq. Yd. Per Unit	Square Yards
Pants	96,111,000	2.4	230,666,000
Shirts	112,119,000	2.7	302,722,000
Long Lab Coats	69,230,000	4.2	290,766,000
Aprons	446,720,000	1.5	670,080,000
		Total	1,494,234,000

Anticipating that in these four categories the durable rental items would become one-time -use disposables, and were sold to industry in place of the rented units, we translated the figures into potential annual consumption for nonwoven disposable yardage. Over one and one half billion square yards of nonwoven disposable goods would be consumed, in only four of the 22 industrial items listed. (See Table 4-3)

THE GRADUATE

Another area of the rental field is graduation caps and gowns. If there is any disposable product in the whole world that can be economically justified, it is an item one wears only once in a lifetime—a graduation outfit.

It is estimated that there are about 770,000 college graduates each year, always increasing. Add to this about two million estimated public high school graduates, plus an estimated 500,000 private school graduates, and one arrives at a a total of 3,270,000. Envisioning only a 10 per cent penetration of this potential market, it is estimated that 1,635,000 square yards of nonwoven fabrics would be consumed annually for disposable commencement gear.

In Cincinnati, school officials claimed the use of disposable outfits eliminated the costly and time-consuming procedures which durable rentals had previously demanded. Students purchased the disposable outfits outright, taking them home as keepsakes. The purchase price (about $5.00) proved to be less than the cost of previous rentals.

Disposables have long been accepted by the American consumer whether she realizes it or not. She has been using disposable paper handkerchiefs for several decades, disposable baby diapers almost as long, throwaway tablecloths and napkins, disposable sanitary napkins disposable wipe cloths, paper plates and cups. In these areas and others, the average American uses over 500 pounds of paper annually. He is already living in a society attuned to the convenience of disposability, and we feel he will readily accept new disposable soft goods items that will further enhance his comfortable way of life. Although few of the products mentioned above can be economically justified, versus the costs of their durable counterparts, they are unquestionably in broad national use and represent big business to the major corporations that market them.

DISPOSABLE DIAPER MARKET

The disposable diaper is a prime example. The market penetration to date is a bare 10 per cent of potential. The full potential is actually the total number of diaper changes in this country every year, which has been estimated at between 23 and 25 billion.

Included among the major corporations presently producing disposable diapers for the consumer market place are Johnson & Johnson, Procter & Gamble, Borden Company, Dennison, International Paper, International Latex, Kendall Company, Parke-Davis, Kimberly-Clark and others. That round-up seems to indicate that big business recognizes the disposable diaper market as a solid-profit opportunity, with 90% of the potential still undeveloped territory.

The disposable diaper leads to many other products. One that is obvious as the next step: disposable baby layettes. What would be a more natural vehicle for complete disposable baby layettes than a "care-for-baby" kit? With about 3.5 million annual births in this country, that's a nice place for disposables to deliver their message to 3.5 million mothers. It is a tailor-made market.

IMPROVEMENTS EVOLVING

Improvements in disposable products are evolving every day. For example, an American Viscose rayon fiber, called RD-101, was specifically developed for certain nonwoven applications. Riegel Textile

TABLE 4-4

Estimated 1967 Production and End Uses For Nonwoven Fabrics

End Use Category	Nonwoven Consumption				
	Dry Process[1]		Wet Process[2]		Total
	Lb.	%	Lb.	%	
Abrasives (pads, discs)	6	100	—	—	6
*(Apparel) (disposable, interlinings,)	19	100	—	—	19
Blankets (needled, padded)	25	100	—	—	25
Carpeting (indoor-outdoor), (needled)	17	100	—	—	17
*Coated Fabrics (laminates, wallcoverings, shoes)	15	94	1	6	16
Filters (milk, air, oil, chemical)	10	53	9	47	19
*Sanitary-Medical (hospital, diaper)	51	81	12	19	63
Tapes & Ribbons (decorative, industrial)	21	100	—	—	21
*Wiping Cloths (dusting, polishing, mopping) (industrial and consumer)	15	88	2	12	17
*Misc. (towels, napkins, drapery, shades, bookbinding, casket liners, aircraft runners)	11	79	3	21	14
Grand Totals:	190	88	27	12	217

(1)—Dry process refers primarily to the textile system for producing nonwovens
(2)—Wet process refers primarily to the paper system for producing nonwovens
*—Disposable end-products are found primarily within these starred categories.

Corporation holds patents covering sanitary products with a nonwoven cover sheet made of this fiber.

One application is a sanitary napkin the performance of which is equal to that of existing sanitary napkin products, until it is introduced to an excess amount of water. The new sanitary napkin, announced by Riegel in 1968, is completly flushable. Water turbulence immediately disintegrates, or disperses, the entire product with no danger of clogging sewage systems. Table 4-4 shows the estimated 1967 production and end uses for nonwoven fabrics-

Comprehensive list of some typical end uses for nonwoven substrates

Abrasive cloth
Abrasive wheels
Absorbent applique
Acoustical materials
Adhesive tape
Advertising novelties
Aprons
Artificial flowers
Artificial leather
Automotive:
    Door panel plumpers
    Headliners
    Quilting and padding
    Upholstery laminates

Bagging fabrics for plastics
Bagging (tea, food, perfume, industrial)
Bandages
Bedpan covers
Bedspreads
Bibs
Book cloth
Bowling towels
Bras
Buffing wheels

Cable wrapping
Camouflage cloth
Capes
Card table covers
Carpet and rug backing
Casket liners
Chafer
Chair bottom cloth
Chamois

Cheese wrap
Clothing insulation
Coated fabric nacking
Coffee bags
Comforters
Cords
Costumes
Coveralls
Curtains
Cushioning materials
Cutoff wheels

Decontamination clothing
Delicate instrument pads
Dental bibs
Diapers—diaper liners
Dish cloths
Doilies
Doll clothes
Drapes
Dresses
Dust cloths

Electrical tape
Facial tissues
Filters:
    Acetate
    Air
    Air conditioning
    Beer and wine
    Chemical
    Coffee
    Dust
    Furnace
    Gas

Milk
Oil
Paint
Filtration felts
Floor backing
Flotation padding
Flower seed rolls
Freezer wrap
Friction tape
Fruit and produce pads
Furniture padding

Garment bags
Gasket material
Gauze
Gift wrapping
Glass wipers
Golf towels
Grass mat
Greeting cards

Handkerchiefs
Hats, (protective type)
Head rests
Holiday decorations
Hospital pads

Ice cream containers
Industrial floor cleaners
Industrial garments
Insulation
Interlinings
    Coats
    Dresses
    Insulated underwear
    Shirts
    Suits
    Ties
Ironing board pads

Jewelry case materials

Labels
Laminated plastics
Lamp shades
Lens tissue
Light diffusers
Liner fabric
Linings for collars, cuffs, belts, girdles
    handbags, luggage

Linoleum manufacture
Lubrication belts
Luggage stiffeners

Map backing
Marking tape
Masks
Mattress covers
Meat wrappers
Medical supplies
Molded and shaped articles
Napkins: luncheon,
    cocktail, dinner,
    commercial

Nursing pads

Oilcloth backing
Operating room covers
Optical lapping wheels

Packaging materials
Parade floats
Pads
Party hats
Pennants and banners
Permanent wave pads
Petticoats
Pillow slips, stuffing and ticking
Pipe coverings
Place mats
Plumpers
Point of pressure pads
Polishing cloth
Protective clothing

Quilting

Radio and TV grill covers
Reinforced plastics
Ribbons

Sanitary napkin cover and pads
Scouring pads
Seed bed covers and carriers
Sheets
Shirts
Shoe goods:
    Doublers
    Fabric
    Innersoles

Liners
Plumpers
Shoe polishing cloths
Shoulder pads
Shrouds
Skirts
Sleeping bags
Sound insulation
Sponges
Stage props
Stiffening fabrics
Surface protectors
Surgical dressings
Survey markers

Tablecloths
Tags
Tailors' patterns

Tampons
Tapes
Ties
Tobacco cloth
Towelling
Tray liners

Underlays
Undergarments
Upholstery backing

Wall coverings and panelings
Wash cloths
Window dressings
Window shades
Wiping cloths
Wire wrapping
Wound cleaners
Wrapping material

Producers of Nonwoven Fabrics in The United States

Company	Plant location
American Felt Co.	Glenville, Conn.
Behr-Manning	Troy, New York
Bender Bros.	Camden, N.J.
Burkart Manufacturing	St. Louis, Mo.
	Detroit, Mich.
Camden Fiber Mills	Philadelphia
Carborundum Corp	Niagara Falls, N.Y.
Carlee Corp.	Rockleigh, N.J.
Carolina Bagging Co.	Henderson, N.C.
Chicopee Manufacturing Corp.	Milltown, N.J.
	North Little Rock, Ark.
	Bensonville, Ill.
Dexter Paper Co.	Windsor Locks, Conn.
E.I. du Pont de Nemours and Co.	Wilmington, Dela.
Felters Co.	Millbury, Mass
Fiberbond Corp. div. Union Carbide	Chicago, Ill.
Fiberglass Industries	Amsterdam, N.Y.
Fiberwoven Co.	New York City
Gustin-Bacon	Kansas City, Mo.
Hollingsworth & Vose	East Walpole, Mass.
International Paper Co.	New York City

Company	Location
Kendall Mills	Walpole, Mass
Kimberly-Clark Corp.	Neenah, Wisconsin
Kimberly-Stevens Corp.	New Milford, Conn.
Loren Products	Lowell, Mass.
Ludlow Manufacturing Co,	Needham Heights, Mass.
Lowndes Products	Philadelphia and Easlet, S.C.
Microtron	Charlotte, N.C.
3M Company	St. Paul, Minn.
National Felt	Easthampton, Mass.
Owens-Corning-Fiberglas	Huntington, Pa.
Pellon Corp., subsidary of Carl Freudenberg Company	Lowell Mass, and Weinheim/ Bergstr. West Germany
Raybestos-Manhattan	Manheim, Pa.
Rogers Corp.	Rogers, Conn.
Sackner Products	Grand Rapids, Mich.
Schwartz Manufacturing Co.	Two Rivers, Wisc.
Scott Paper Co.	Chester, Pa.
Star Textile & Research, div. Allen Industries	Cohoes, N.Y.
Stearns & Foster	Lockland, Ohio
St. Regis Paper Co.	Deferiet, N.Y.
Troy Mills	Troy, N.Y.
Union Wadding Co.	Pawtucket, R.I.
United Cotton Products	Fall River, Mass.
U.S. Rubber Co.	Hogansville, Ga.
West Point Pepperill Manufacturing Co.	West Point, Ga.
West Virginia Pulp and Paper Co.	New York, N.Y.
Wood Conversion Co.	Cloquet, Minn.

Typical Nonwoven Fabric Producers Outside the United States

Country	Company	Plant
Canada	Dominion Rubber Co.	Kitchener, Ontario
	Montreal Quilting	Montreal, Que.
England	Bondina Ltd.	Greetland, Yorkshire
	Bonded Fibers	Bridgewater, Somerset

Country	Company	Plant
	J.B. Broadley	Rawenstall, Lancashire
	Johnson Fabrics Ltd.	Earby, Colne, Lancashire
	ICI Fibers	Harrogate, Yorkshire
	Lansil	Lancaster
	Luxan Ltd.	Oldham, Lancashire
	Dunlop	London, England
	Robert Pickles Ltd.	Burnley, Lancashire
	Smith & Nephew Ltd.	Welwyn Garden City, Hertfordshire
	Southalls Ltd.	Birmingham
	Tootal Broadhurst Lee	Manchester
Scotland	Jute Industries Ltd.	Dundee
Ireland	Irish Ropes	Newbridge
France	Ferodo	Conde-Sur-Noireau(Orne)
	Intissel, S.A.	Wattrelos (Nord)
	Coisne & Lambert	Armentieres (Nord)
	Motte-Bossut	Roubaix (Nord)
	Est. St. Denis	Brionne (Eure)
	Est. Mulsant-Rouches	Villefranche-Sur-Saone
	Mfg. De Feutres De Mouzon	Mouzon (Ardennes)
	Comptoir Linier	Paris
West Germany	Carl Fruedenberg	Weinheim/Bergstr.
	Karl Lissman, K.G.	Muchen-Sulin
	Vereinigte Glanzstoff	Wuppertal Eberfeld
	Vliesena	Essen
	Feldmuehle	Dusseldorf
	Schickedanz	Nuremberg
	Filzfabrik Fulda	Fulda
	Jute Webberie	Emsdetten
	Lohmann KG	Fahr (Rhein)
	Woll-Wascherei Und Kammerei	Hannover
	Textilana	Giesen
Italy	Manufactura Valle-Olona	Castellanza, Milano
	Filatura DiLana Pettinata	Cassano Magnag, Varese
	Tenotex	Milano
	Cia Viscosa	Padua
Holland	Hollanter N.V.	Veenedaal
Switzerland	Grossman & CO., Ltd.	Thalwill

Country	Company	Plant
Mexico	Milyon S.A.	Mexico City
Australia	Felton Textiles	Melbourne
	Lantor of Australia	Devenport, Tasmania
Japan	Nippon Cloth Co.	Nishikyogoku, Kyoto
	Fuji Hat Mfg. Co.	
	Kanai Juyokogyo Co.	
	Japan Felt Industrial Co.	
	Dai Nippon Spinning Co.	
	Tokyo Acetate Co.	
	Kurashiki Rayon Co.	Osaka
	Ritto Sen-1 Co.	
	Japan Vilene Co.	Gifucity
	(Toyo Rayon)	
	(Dai Nippon Ink)	
	(Carl Freudenberg)	
	Fuji Spinning Co.	Fujibo
	Nippon Keori	
	Teijin Nonwoven Co. Tokyo	
	Kureha Spinning Co. Osaka	
	Kanekaon	Osaka
	Toyo Spinning Co.	Osaka
	Nippon Keori	
	Tsuyakin Industry	Nagoya
	Kinsei Seishi	(Kochi) Shikoku Island
	Pulp Co	,, ,, ,,
Argentina	Suilene Argentina	Buenos Aires
Finland	AB Avatex Oy	Hammarland, Aland
Sweden	Molnlycke A.B.	Gothenburg

Note:
Several firms in the United States and Europe are planning to enter the non-woven field in the coming year with processes that vary from tissue fiber laminates to wet process, dry process, needled, and spunbonded and to more sophisticated systems. Because these new sophisticated techniques are not commercialized to date, we refrain from mentioning them.

Special note to add:

Austria	Bunzl-Biach	Vienna
New Zealand	Arthur Ellis Co.	Dunedin

SIGNIFICANT TRENDS AND EMERGING TECHNOLOGIES

World-wide research is exploring the profitable new products that can be manufactured with sophisticated new systems and processes. Breathable films will play an important part in the development of nonwovens. Coupled to this, film fibrillation to manufacture new web structures will expand the uses of nonwovens.

Major fiber, chemical, paper, and textile manufacturers will allot additional monies for research and development in hope of producing a more economical nonwoven or a substrate with "unique" properties.

Clearly in today's atmosphere, one can assuredly expect new breakthroughs in exotic spunbonded materials where the core of the fiber may be of different polymeric construction than the sheath to create new improved properties not now obtainable with nonwovens.

Every major fiber company sees new potential profitable products by applying and extending their technology of fiber production to new spunbonded systems.

Paper companies and fiber companies will probably play the dominant role in the near future commercialization of new products. However, to be successful in the manufacture of finished products, they will need the creative skills of machinery manufacturers to develop new systems of converting the nonwovens to end products in the most economical manner.

MARKETING STRATEGY

We have examined the characteristic parameters of technically intensive businesses and their impact on strategic, administrative, and operating problems of top managers. By way of summary, let us consider the impact of these characteristics on a strategic issue: the timing of the technologically intensive firm's entry into an emerging industry. The alternatives may usefully be grouped into four major marketing strategies, recognizing that most companies will—or should —adopt a blend of these according to the requirements of their different markets or product lines:

First to market—based on a strong R & D program, technical leadership, and risk taking

Follow the leader—based on strong development resources and an ability to react quickly as the market starts its growth phase

Application engineering—based on product modifications to fit needs of particular customers in a mature market

"*Me-too*"—based on superior manufacturing efficiency and cost control.

The manufacture of nonwovens or disposable substrates is a technology-based business, and therefore requires a careful analysis of the direction and time a company should commit itself to a capital expenditure. The above is a guideline, but additional answers must be obtained to further inquires about 1970 or even 1980 like these:

(This paragraph was retyped because of the many confusing corrections.)

TEXTILE REPLACEMENTS AND THEIR FUTURE.

The purpose of this section is to make the various representatives of the textile, paper, chemical, and fiber firms consider the near-term and long-term future in'relation to new materials or substrates so each will play a major role in creating, manufacturing, marketing, and collaborating to assist each other in upgrading this new industry.

(1)   How much plant capacity should we have to meet this probable market of the future?

(2)   What kinds of new products and new production technologies should we have under development to get ready for thefuture?

(3)   What will be the future market demand for our wares?

(4)   How large and how skilled will our employee force have to be?

These of course are straight bread-and-butter questions. There are also broader questions like:

(1)   In what direction and how far will the economy move in the years ahead—up, down, or sideways?

(2)   What will be the upcoming sociopolitical structure of values, mores, and institutions under which business will operate?

Hopefully, the framework for a real interindustry group will be formed and brought into operation to benefit all.

Only through the upgrading of standards and quality will this new industry grow and remain profitable.

Everyone has been so much interested lately in the subject of textile replacements and their future that this may be an opportune time to review what has been accomplished and to determine what we can do to produce new and profitable products to satisfy the demands of the many markets.

We have in the United States and abroad, textile, paper and chemical firms that have a vital stake in textile replacements. Each views the challenge from a different discipline, but the key to the future for success will depend upon the collaboration of all the groups within the industries. The resultant of this effort will be to:

(a) Give a cross fertilization of ideas.
(b) Upgrade the quality of materials produced.
(c) Upgrade the specifications.
(d) Create more acceptable products for the consumer and industrial markets.
(e) Create an understanding so that top management will support new research programs.

The modest success to date and the fact that the nonwoven industry has done a good job under great opposition from the many forces such as competition from textile mills, lack of acceptance from manufacturers that are afraid to use advanced technology, coupled to the task of educating the various markets to the applications and improved performance of nonwovens is an inspiration to the future profitable opportunities that lie ahead.

There have been some drastic changes going on with the entry of new substrates researched and developed by key factors in the paper, chemical, and fiber industry. This surely changes the character of the nonwoven industry with the entry of new concepts, new products, coupled to advanced technology.

The textile industry today recognizes that there is a paper industry. In addition, the nonwoven manufacturers are starting to tell their prospective customers that their material is a replacement in fact. For many years the mills avoided the use of this word, but the advance and acceptance of paper substrates and other substrates as a direct replacement for woven and knitted materials has actually upgraded conventional nonwovens.

The textile process has the main part of this new and growing market, but in the future the chemical and paper companies may have the dominant role. The textile industry must not be afraid of the paper industry. Textile people and paper people should pull together because the "Disposable" concept is coming—like it or not.

Textile replacements have been around for some time; however, industrial demands for new substrates have sparked a crash program resulting in the acceptance of new textile replacements. To cite an industrial example, let us examine the hospital situation briefly. For years, rising costs, labor, materials, and the harassment of the staph germ have created a real need for better management and the use of new materials to keep costs down, but also to assist in controlling the spread of the staph germ. First came plastic products for dishes and eventually, a line of disposable needles and syringes. Following this, a need for operating gowns and their components was recognized, not only for convenience of the use of the product, but also because it assisted in the operating procedure and possibly cut down on staph infection through the use of the discardable material. Here is an illustration where the paper material or substrate is a direct replacement of textile materials. The significance of this is easily illustrated. The next stage of development is other hospital products that eventually will find the way to other industrial uniforms and then to consumer products at a later date. New minimum wage laws and Medicare result in hospitals using modern cost accounting. This will work in favor of disposables. Let us now take a look at the background of textile replacements.

History of Nonwovens

Nonwovens have been manufactured for many years but have been sold primarily on a reciprocity basis. The total volume was small and was consumed primarily by parent companies through subsidiaries. When World War II broke out, there was a shortage of lining materials, and those mills that produced nonwovens found a demand for their goods.

When the Carl Freudenberg Company of Germany established American operations about 1951, they had already created through research and development and manufactured and sold improved nonwoven lining materials. However, tremendous sums of money were set aside to develop replacement materials for other industrial applications. One was the shoe trade, where Freudenberg had a vested interest as the largest tanner in western Europe. They developed and sold in the United States vast quantities of shoe liners and innersoles.

At the same time the American mills were alerted to the success of their foreign competition, they decided to develop new and perfected substrates. However, there were other needs outside the conventional nonwovens that the American mills were not fulfilling.

In the years 1951 to 1969, great strides were made by the producers of nonwovens. However, few if any actually approached the market and called their material a "replacement." Wonderful work and progress were made in the interlining trades, automotive trim backing, shoe industry, air filtration, and synthetic leather backings.

At this very same time, paper firms, under patents, were improving the technology of providing a nonwoven scrim base for the production of disposable or discardable paper-like materials. In 1957 when Kimberly-Clark showed the Jinx Falkenberg tennis dress in an exploratory test, all the nonwoven people said that this could not be a threat to their future. They virtually ignored this new technology until they saw the beginning of the new era in the adoption, maybe slow at the start, but increasing in use in the hospital industry.

An interesting example whereby collaboration between two companies is working out fairly well is the case of operators of a nonwoven mill who did not develop a satisfactory hospital substrate but in turn are buying time, so to speak, by purchasing a competitor's substrate until they can perfect their own. In the meantime they are continuing their efforts in this fast-growing and profitable hospital market while they are perfecting their own material. They are also keeping up on the state of the art and making money. It might be that they will be able to use their own and purchased material in their vast product line at a later date. In any event the substrates employed are a direct replacement of textile materials.

In 1959 special work was done to explore fibers and fibrids to produce very interesting and potentially profitable substrates on modified paper machinery. Tremendous sums of money, manpower, research and development, and marketing were employed to open up new markets that would be a direct replacement of conventional textiles. Unfortunately, after a few years of struggle, the firms engaged in this technique resigned from the project, and it was found to be more economical for the fiber company to produce the substrate directly with their new technique, resulting in spunbonded materials.

Another phase of technology was the introduction of needle-punched nonwovens. This added a new dimension and improved the quality of those substrates that were employed as a backing on automobile upholstery. This has been enlarged by the introduction of nonwoven or needled blankets and outdoor-indoor carpeting.

All the methods of producing materials, whether the wet or dry processes, needled, tissue-impregnated, scrim-reinforced, or made by spreading cigarette tow on films require the knowledge and expertise of the backgrounds represented. There are very definite areas of collaboration that can eventuate by cooperation.

TECHNICAL RESEARCH AND DEVELOPMENT

The importance of research and development cannot be stressed enough. The trend today is for research and development's responsibility for new product profitability tomorrow. The R & D program can make or break a company. Success rests largely on the ability of the research director and his associates to recognize or to conceive suitable lines of inquiry and to grasp or to generate significant opportunities in advance of less alert and nimble competition. This is especially true of that part of the program dedicated to the quest for new products that are to broaden markets and to upgrade and expand profits significantly over periods measured in years, and where passage from idea to marketplace may call for extensive endeavor, a major technological advance, and substantial capital outlays. The company must also recognize new trends in an economic society, i.e., the automotive society and now the disposable society.

In the hunt for appropriate projects, sales people seldom bring in big game. Their special talents find exercise elsewhere: courting new customers, caring for old ones, fending off competition, and pressuring management for higher quality at lower prices.

Market research people and marketing people are similarly busy doing what they do best: identifying economic trends, planning and conducting opinion polls, garnering trade data and skillfully extracting valuable information from many sources.

Production people are likewise concerned about their immediate problems: safety, labor costs, waste reduction, cheaper raw materials, process changes that would permit wider latitude in manufacturing procedures and the like.

Production people should not only be concerned with having their machines run faster, but should develop flexibility in their use. Committees and groups of uninhibited brainstormers, sometimes, are potentially valuable sources of superior concepts.

R & D people know best, or should know best, where their company stands relative to the state of pertinent technology. They therefore, are best prepared to bear responsibility for the generation of new product profitability. Of course the sharp company has "the better part of two worlds:" good marketing and good research. One good one without the other is not sufficient! In support of this view, consider a few condensed case histories, of which firsthand information is available.

The first took place about 1952. The basic and exceedingly effective plan of the Carl Freudenberg Company in Germany was to do everything possible to maintain its products at the highest practical level of quality. Profits were excellent and growing. However, during the prior year their wholly-owned and newly-established counterpart in America, Pellon Corporation, was having great difficulty introducing those non-wovens that were already established in Germany. The fashion market at about this time was creating flared skirts with cotton fabrics that were fully starched to create the flare, but petticoats were lacking in performance! A crash program was started in Freudenberg's laboratory, and by the end of the year when the full flared skirt was really big, Freudenberg gave their counterpart in America a compound bonding agent that increased the resiliency and actually improved the petticoat construction. This was a direct replacement of cotton with synthetic woven materials. This business grew so rapidly that some customers were on an allotment basis. It also gave Pellon its first profitable year. These profits were carefully set aside to finance further research and development of new products.

The second case occurred about 1957. Lining materials of nonwoven structure were facing great competition from woven fabrics. There was a compelling need to perform better than could woven materials, especially because in some instances, the nonwoven substrates cost more. The profit obtained at that time from the sale of nonwovens was required because of the expenses involved for technical sales assistance, product development, and the high cost of the rubber bonding agent employed.

At about this time a small, but technically oriented, English firm was experiencing excellent success from the sale of fusing materials in Europe and elsewhere. Some of their product was then sold in the United States and R & D saw an opportunity to develop nonwoven fusing materials for the fashion trade.

Now it should be relatively easy to demonstrate the effectiveness of a fusing material over a standard lining. A search was made by Freudenberg in Europe, where at that time the research was carried out, to locate the proper blend of polyethylene powder to dust the surface of the substrate. In cooperation with firms like Billeter and Schaetti of Zurich, I. C. I. and Liquid Nitrogen Processing Company in Philadelphia, the proper blend of Coathylene and Lupolene was made and then pulverized on a Pallmann mill prior to dusting the web. This is history today, because the race is on for a wide variety of fusing materials to replace woven fabrics. There is a significant trend to adopt fused parts to save time and money in the manufacture of garments. This is an area of real growth.

The third case dates from about 1956. Research & Development worked on the problem of developing a nonwoven substrate that had certain characteristics of stretch and pliability. In time, R & D came up with the solution which opened the door to new markets and helped increase business in existing areas. Up to this date no mill had commercially produced stretchable nonwoven materials.

The fourth case occurred about 1959. Kimberly-Clark researched and designed machinery to produce what is known as Kaycel. However, at about the same time, a firm known as American Sisal-Kraft was also working in this direction, but was considering the heavy scrims and not the light ones of interest to Kimberly-Clark.

Subsequently, St. Regis Paper Company purchased American Sisal-kraft and today occupies a strategic position in this new paper business, because it controls patented processes, which are basic to both Scott Paper's Duraweve and Kimberly-Clark's Kaycel. Both materials are made of loose rayon or nylon mesh called scrim, which is bonded to several layers of wood cellulose wadding. In the recent years Kimberly-Clark has made tremendous strides in updating its technology and is producing Kimlon, which is constructed differently from Kaycel and has a different technology, not covered by St. Regis Patents. All these materials are textile replacements. This is an illustration where marketing worked hand-in-hand with R & D.

There are many more cases that one could cite, such as the development of Corfam, where R & D came up with the answer to the need for a new material. Although synthetic leather has been produced for many years, it was Du Pont that saw the vast market potential coupled to the growing shortage of animal skins, to commercially develop and market Corfam. There is not sufficient space here to elaborate on the many other fine R & D projects that were brought to completion.

Having presented four case histories of successful planning and development of profitable new products, one should call attention to the significant features common to all four:

(1) Each project was conceived in the research department, most with direction from market research people. Success abounds when one complements the other.

(2) Each was carried through all stages from research to development to early production runs to sales service by or under the direct supervision of research personnel; in other words, the project was protected by a determined and often combative research department against disappearance in other departments.

(3) Each project was technologically and otherwise extremely difficult

to carry to a successful conclusion and therefore of limited appeal to others less venturesome.

(4) Recognition of the opportunity that underlay each project was mainly a creative exercise

(5) In each case the product to be developed, or its intended use, could be accurately defined.

(6) Each project was based on the concept of a product for which at least one important use was assured from the outset.

(7) In each case the envisioned product or family of products fitted nicely into one or another of the corporation's specialized capabilities.

(8) In connection with each project, top management was steadfast in its support.

To venture a formula for success in product innovation on the basis of the foregoing list of features would be folly indeed. No obvious formula exists for success in any complex human undertaking in which creativity plays a large part. One could cite many cases of projects propitiously started and scientifically implemented that came to naught. Every project of substantial magnitude aimed at high stakes is threatened by uncertainties, for example, an anticipatory innovation, a patent snag, a ruinous change in price structure, a technical impasse, a shifting fashion trend. However, it seems reasonable to suggest that if a project devoted to product innovation of major scope has features similar to those summarized for the cases cited, and if the quality of research leadership is of proven merit, fortune will favor the sponsor. Expressed in management terms, the conditions would then be optimum for the generation of new product profitability.

GATHERING COMPETITIVE INTELLIGENCE

Some firms shy away from the use of the word, "intelligence," but the Vice President of American Bosch Arma, speaking recently, said "Let's call it what it really is. There is nothing wrong with the word, and most firms are employing this technique in one form or another."

The real problem is that too many medium and small firms do not use market research and do not gather sufficient competitive data in weighing some of the decisions they have to make in the face of competition. Naturally, there are proper ways and some that are questionable. Every firm listed in Fortune's 500 top companies has intelligence organizations or counterintelligence organizations which they call market research or market development. In addition, key firms from overseas employ firms in the United States to keep abreast of the latest raw materials and products produced that are or might be in competition with their firms.

MARKETING INTELLIGENCE

The traditional notion of marketing research is fast becoming antiquated for it can lead to dreary chronicles of the past, rather than focusing on the present and shedding light on the future. It is particularistic, often tending to concentrate on the study of tiny fractions of a marketing problem rather than on the problem as a whole. It lends itself to assuaging the curiosity of the moment, to fire-fighting, to resolving internecine disputes.

Each firm should have a marketing intelligence system tailored to its specific needs. Such a system would serve as the ever-alert nerve center of the marketing operation. It would have these major characteristics:

(a) A variety of research techniques
(b) Continuous surveillance of the market
(c) A network of data sources
(d) Integrated analysis of data from the various sources
(e) Effective utilization of automatic data-processing equipment to speedily concentrate mountains of raw information
(f) Strong direction, not just on reporting findings, but also on practical, action-oriented recommendations.

BENEFITS REALIZED

In addition to giving executives early warning of new trends and problems and valuable insights into future conditions, it leads to a systematic knowledge of company markets rather than to isolated scraps of information. This knowledge in turn should lead ultimately to a theory of marketing in each field that will explain the mysteries that baffle marketers today. What is more, a company will find that the system will help to free its marketing intelligence people fromfire-fighting projects so that they can concentrate on long-term factors and eventually be more consistently creative. We require more marketing intelligence. We should each examine our own system to improve it, and, more important, we should not have a marketing research department for prestige, but management must realize that this function is an intelligence tool for the marketplace and for the future.

In summarizing the gathering of market intelligence, it is an entirely proper function of any company. In fact, it is inconceivable that any company that wants to stay in business and grow can play the ostrich and "keep its head in the sand."

We must know what's going on. We must know such things as market trends, needs, developments, attitudes, economics, and growth projections. The following are the key points to remember in gathering marketing intelligence:

(1) Make a thorough market investigation;
(2) Enlist the entire field force;
(3) Gather data from distributors;
(4) Use shows and conventions;
(4) Follow the financial and business publications;
(6) Consult trade and government publications;
(7) Check competitive catalogs and promotional material;
(8) Check the competitive product, where possible.

All the foregoing activities are proper. Correct evaluation or interpretation of the intelligence is the key. Information is just information —just a part of the over-all picture—every item, every bit of information, must be correlated, sifted, put together, compared, projected, and appraised. Only by such a thorough evaluation of all the intelligence gleaned can a correct course of action be determined.

TEXTILE REPLACEMENTS AND THEIR FUTURE

Paper Background

In 1919 the per capita consumption of paper in the United States was 120 pounds; today it is 530 pounds. Many types of paper products that were nonexistent or rare at the end of the first world war are nowadays commonplace in the household or in the office; for example, absorbent towels and napkins that hold together when wet, facial tissue, office copy paper activated by heat rays, relatively inexpensive bright white opaque text papers unusually light in weight suitable for high speed lithographic printing, the ubiquitous milk carton, sanitary napkins, and many others. Among the newer papers developed for industrial use are oil and air filters, inexpensive graphic arts papers coated at high speed on the papermaking machine, sandpaper bases and photographic printing papers manufactured from newer types of wood pulp, paper base artificial leather insoling for women's shoes, and cigarette paper manufactured from seed flax tow. Whereas in the recent past most paper was manufactured from coniferous trees, research has taught us how to convert a wide variety of deciduous trees into valuable pulp and paper products. The modern rayon and cellophane industries are based on purified wood cellulose. During the second world war specially prepared wood pulp for the first time in history was used on a huge scale in the manufacture of propellant powder. Improvements in forestry practice, in pulping processes, in papermaking, in our understanding of the nature of wood, the development of numerous and valuable paper additives and other advances have contributed enormously to the phenomenal growth in size and resources of the industry.

Sanitary and Medical Field

This has been the single largest growth area for textile replacements. Papers, films, reinforced papers and combinations are being employed in this field. A spectacular breakthrough has been accomplished in the past years with the adoption of scrim reinforced papers for use in surgical packs and other related items. Chicopee (Johnson and Johnson) and Kendall were among the earliest producers of nonwovens and they concentrated their early attention on these applications because they were in a strong marketing position and because they recognized the vast opportunity and potential for new profits in the sanitary and medical field. In 1959 firm emphasis was placed on the commercialization of a surgical pack known as an O. B. (obstetrical) and O. R. (operating) pack that would contain all the vital elements for the operating room and one that could be presterilized. Six years later after overcoming many technical difficulties, acceptance of the surgical pack was felt. Market penetration is estimated as 5%, consuming over 25 million 44-inch yards annually. This spurred on a crash program by other firms that recognized all the time the importance of the surgical pack concept but were unable to produce a suitable material in their own mills.

Through cooperation of some firms, important substrates were adopted and expanded plant facilities were provided to promote the surgical packs.

This field is so important for future growth that management at key firms has established crash programs to enter this very important and profitable field. *The significance of this illustration* is that in the future the mills that produce the basic substrates will have to take a more important role in converting their materials to finished products. At the same time companies in the hospital supply business will be pressed to find an equity position with paper mills. *The second point* in this illustration is that the hospitals are benefiting from improved performance of surgical packs from more than one source, with high specifications set and quality maintained.

Standard nonwovens and films will also play an important role in creating additional sanitary and medical products.

In short, the hospital field is now ready and will welcome the new textile replacements. The United States has about 6900 hospitals with a capacity of about 1,658,000 beds and an average occupancy of 84.6 per cent, taking care of 500 million patient-days a year. With an average use of 10 pounds of linen each patient-day, the total volume of linen processed in hospitals each year is 5 billion pounds. This market also presents some key problems as to the manner in which the hospitals will eliminate their disposable materials after use.

Fashion Field

Much has been written on the subject; it is timely but not significant.

Apparel Interlinings

In the next twenty years virtually every garment will be interlined with a nonwoven substrate. Little or no woven haircloth will be used.

Carpet Industry

The new dimension of needled felts and nonwoven carpets has created a "new carpet" industry. In the near future we will see surface effects that will be hard to distinguish from conventional carpets.

Shoe Industry

We have all witnessed the slow but steady growth of nonwoven lining components used here to replace cotton fabrics. We also have available new polymeric lining materials.

Coated Fabrics

An early replacement of a textile in the automotive industry was a nonwoven replacing cotton knits.

Miscellaneous

Dusting, polishing, shining, mopping, buffing, and wiping, are a few items into which nonwovens have made great inroads and will so continue. Nonwovens will gradually replace virtually all conventional textiles in the hospital industry and many in industrial applications. Their growth will be limited only by the lack of creative work to develop newer concepts in manufacturing and in marketing.

CONCLUSIONS

To summarize some of the key points and look at the near term future:

(1) The future looks very bright for nonwovens, but the many problems to be solved will require more collaboration from the inter-industry groups.

(2) New inter-industry groups should be formed.

(3) Bigger budgets will be given to research and development.

(4) The trend today is for increased responsibility for new product profitability on the part of research and development.

(5) More emphasis will be given to the use of market research and the gathering of competitive intelligence.

(6) Many combinations of crossbreeds will appear on the market, for example:
    —Scrim reinforced films,
    —Nonwoven scrims laminated to tissue,
    —Woven scrims laminated to tissue,
    —Long staple fibers laminated to tissue and films,
    —Cigarette tow spread on films and paper, for textile replacements.

(7) A new "garment" industry is springing up away from Avenue to convert these many new substrates for industrial and hospital products.

(8) Large paper mills will necessarily convert those highly technical products where extreme quality and high standards exist.

(9) Paper firms will, by necessity to reach other industrial markets, have subcontractors convert their products.

(10) Nonwoven mills that previously sold on a reciprocity basis to parent firms will play a bigger part in supporting the hospital industry's demands.

(11) The consolidated laundry linen supply companies will play an important part in producing and distributing industrial uniforms.

(12) All hospital supply companies will play an active role in converting to produce their own products and as the market grows, will probably acquire their own paper mills.

(13) Fusing webs made from new, exotic polymers will be employed to eliminate sewing.

(14) Certain preformed products will be produced directly from paper machines.
(15) New materials-handling machinery will aid in cutting the cost of converting.
(16) We shall see a host of new spunbonded substrates coming from the leading fiber and chemical firms.
(17) We shall see the use of water-soluble films and specialty fibers to produce flushable products in both the medical and sanitary fields. These are direct replacements of woven textiles.

chapter 5

Extrusion Coating of
Paper and Board

Extrusion coating of polyethylene and other thermoplastics has been a commerical art since about 1948. As with many commercial arts, interest and enthusiasm outpaced technology in the earlier stages of growth, and much of today's know-how was developed out of engineering, astrology, physics, curses, chemistry, and applied black magic. Sorcery is no longer a necessity in running an extrusion coating line, because of diligent trial work on commercial equipment and some heroic efforts on the part of resin suppliers, all goaded by a need for a uniform, predictable product.

SURVEY

Although many in the paper industry view plastics as a threatening competitive product,[1,2] the extrusion coating process sold approximately 158,000 tons of polyethylene plastic coated to about 1,250,000 tons of paper and paperboard in 1967, most of it because of functional or decorative properties that neither a paper nor plastic product could develop by itself for equal cost. No single set of properties brought this about. Both a paper substrate and a polyethylene coating or lamination can be tailored for a specific character according to specific design requirements. Available paper grades and resin types can be combined to take advantage of combination of the following properties, although no single grade of paper and no single plastic resin has all of the properties listed:

Paper		Polyethylene
1. Opacity	A.	Transparency

2.	Stiffness	B.	Vapor barrier
3.	Print surface	C.	Heat-sealability
4.	Moldability	D.	Opacity
5.	Greaseproofness	E.	Washability
6.	Glueability	F.	Flexibility
7.	Transparency	G.	Smoothness
8.	Flexibility	H.	Slip
9.	Wide color range	I.	Water-shedding
10.	Physical strength	J.	Release
11.	Absorbency	K.	Chemical inertness
12.	Dead fold	L.	Gloss
		M.	Print surface
		N.	Grease and oil barrier

The list is illustrative if incomplete. For example, a manufacturer of machinery wants to package small spare parts, inventoried in units of eight pieces. One satisfactory design for a package would be a transparent glassine coated with clear polyethylene. Properties utilized would be 3, 5, 7, 8, A, C and N. Another case may be a jewelry manufacturer, who requires a decorative cover for his sample display boxes. The grade selected may be a printed bleached kraft paper, polycoated over the print copy, and embossed in a leather pattern. This application would take advantage of 1, 3, 4, 6, 9, 12, A, E, and L. Both substrate and coating are chosen to develop the necessary attributes in the finished combination.

Although the extrusion coating method is not limited to polyethylene, "extrusion coating" and "polycoating" are almost synonymous. Some definite advantages brought polyethylene into early use and today's predominance as coating resin. Polyethylene is inexpensive and is quite uniform in processing properties through successive purchases.

Additionally, polyethylene has a low density for a plastic, 0.914-0.960 gm/cc, offering a "mileage" advantage in functional coatings. The resin needs no plasticizing or other elaborate compounding for most of its coating applications, and a wide range of resins is continuously available, including a variety of colors.

Polyethylene-coated and laminated papers have penetrated many substantial markets; often it would be more precise to say that a market appeared when a line of products became available to fill its needs. Some of the major sales items of extrusion coated paper and board follow.

Milk Cartons

Including cream, orange juice, and other fruit drinks. Replacing an earlier wax-coated paperboard structure, a more or less standard two-side polycoated carton accounts for the largest single use of coating-grade polyethylene resin. The present grade is usually flexographically printed over a treated* polymer surface, although there has been some real interest in web-offset decoration for added sales appeal.

Multiwall Bags

One or more plies of bagmaking kraft paper are coated with poly-ethylene for moisture protection of the contents or to prevent escape of dust or oil. Some added puncture protection is also gained. Bag paper coating and the foregoing milk carton coating have accounted for the installation of some really large (105-132 inches wide) coating lines in the South.

Composite Cans

Principally for motor and other petroleum oils. In 1968 this item was less than five years old, but already accounts for considerable tonnage of MG kraft paper and high-density polyethylene. Motor oil cans of composite fiber spiral construction have a barrier layer of light-weight (20-25 lb/3M Ft^2) kraft paper coated with 10-15 lb of high density polymer, which are not attacked by even quite warm lubricating agents. Use of composite cans will probably continue to increase as grades are worked out for other items to be packaged, such as anti-freeze (for which some polypropylene coated paper is already being used), brake fluid, possibly vegetable shortening, coffee, and other food items.

General Food Packaging

–divided into a number of major areas

Frozen food locker paper

–a colored or white grade coated with 5-10 lb of polyethylene for a moisture barrier and easy release from frozen items. Some grades may be heat-sealed.

---

* Flame or electric corona treated for ink adhesion.

Bakery and candy wrap

–often coated on printed paper for product identity, offering moisture and grease protection and a heat-seal closure.

Heat-sealing pouches

–usually of the "form-and-fill" type, being assembled in the packaging machine from roll stock, then filled and sealed before leaving the machine. Multilayer laminated combinations for pouches often contain plastic films or aluminum foil, the latter being a virtually complete vapor and gas barrier. Foil also excludes light, a real factor in flavor protection of certain foods. Types of structure and requirements for them vary greatly according to the product and the machine anticipated for the package-making operation. At prsent, form-and-fill packaging is commonly used for packaging of

Individual sugar portions (plus non-caloric sweeteners and coffee lightener)
Cake and other baking mixes
Soft drink concentrates (often a foil-laminated construction for flavor retention)
Powdered milk in small-use packets
Frozen confections, such as ice cream bars
Instant breakfast cereals, in single portions
Prepared cocktail mixes
Instant potato products.

"Industrial" food grades

–substantial tonnages of polymer-coated papers are used for handling and shipping only and never appear as part of a retail package. Such products include candy drop paper used in making fancy confections, case lining sheets, and interleavers for meats and baked goods.

Tray and cartons

–for smoked meats and bakery, as well as unprinted cartons for frozen foods, to be overwrapped.

Industrial Wrapping

This category covers a range of protective papers, such as interleavers for stacking painted and polished metal shapes, greaseproof wrappers for oiled metallic items, (some incororating rust-inhibiting agents) and heavier laminated and reinforced grades for crate tops and outdoor storage packages.

Construction and Shelter

–heavy-duty papers, usually paperboard-based or coated onto a reinforced paper substrate. Such grades are often coated with polyethylene on both sides and are used for crawl space and foundation vapor barriers, and for walls for temporary shelters and enclosures. Some tonnage is used as a one-time tarpaulin over straw stacks and piles of unshelled corn and other produce.

Decorative Cover Papers

–products may include heatseal overwraps for cartons, facing sheets for retail set-up boxes, covers for phonograph and other hard cases, and low-cost wall and ceiling facing. Plain or printed papers can be used, coated with polyethylene or possibly polyamide or vinyl materials and often embossed to simulate leather or woven patterns.

Many other markets of importance for extrusion coated products also exist. Pouch packaging of household cleaners and chemical products is increasing. Serious interest has been shown recently in a weed-inhibiting mulch paper for agricultural row crops, utilizing a low-cost kraft sheet with a light-weight coating of clear or black polyethylene. Ideally, a season's exposure to the weather would disintegrate the structure sufficiently that the paper can simply be plowed under after harvest.

Some converters have been seriously working on package structures in which a coating of polyethylene becomes a smooth, pinhole-free substrate for some other functional coating, such as silicone release materials or special gas-barrier polymers. (Polyethylene is only a modest barrier against gas and odor penetration.) Although anchoring of overcoatings has been a problem, polyethylene can easily be given a mirror-smooth, level, pinhole-free surface, ideal for developing the best mileage of an overcoating material.

Development of Polyethylene as a Paper Coating

Despite the wide use of polyethylene today, (and all predictions point to substantially greater use tomorrow) it is a relatively new plastic. The technology of additional polymerization was already well developed in the 1930's, with successful production of polystyrene, acrylic plastics, and a wide variety of synthetic rubbers. The polymerization

 UNREACTED
ETHYLENE

 CATALYZED
ACTIVATION

 COMPLETED
POLYMERIZATION

Fig. 5-1. Addition polymerization of ethylene gas to form polyethylene.

of all these involves activation and union of an ethylenic group (-CH::
CH-) in the molecule of the parent monomer.[3] Thus, a reaction might
be as follows: (Fig. 5-1) in which (N) is some large number, over one
hundred, which usually defines a polymer plastic.

Ironically, the simplest of all compunds containing the active ethylenic group, ethylene itself, for a long time resisted efforts at polymerization.[4] The first real "plastic" from ethylene was produced in 1933 by
Fawcett and Gibson[5], who successfully catalyzed an addition reaction
at over 20,000 psi.

Because of experience with related plastics in pilot plant and commercial development, ICI in England was ready for commerical production
of "poly" by 1940. Some very small amounts had been produced before this and were used for underwater cable insulation as a synthetic
replacement for gutta-percha.[6] The material proved to be almost
completely inert to water and dissolved sea salts, and dielectric losses
were small and predictable.

During World War II, gutta-percha and rubber became scarce, and
polyethylene replaced more of these materials as electrical insulation,
first in Great Britain and gradually after 1943[7], in the USA. Only
after the war was there enough production capacity available to allow
a serious look at other useful properties of polyethylene. The end of
the war and expiration of some of the original patents brought this
material to the market place.

Shortly after World War II, polyethylene film became a commercial
item, and rapid progress in technology, a steady decline in the price of
the resin, plus a yield advantage because of the low density of the

plastic combined to bring polyethylene film into dirct competition with cellophane and polystyrene. The idea of extrusion coating onto paper or a similar flat substrate comes easily to anyone watching a film extrusion line in action. The first commercial extrusion coating operation was performed in 1948, by St. Regis Paper Co., principally for multiwall bag barrier paper, and by 1952, at least one 120-inch wide coater was running. Throughout the 1950's, polymer resin, although still expensive by today's point of view, declined steadily in cost, and more and more markets became open to polycoated papers. By 1960, a number of coaters had given up buying their resin in 50-lb paper bags and had installed systems for handling bulk resin in 1,000 palleted containers, large rubber balloon containers holding 10,000 lb, or had installed silos for storing bulk plastic from 100,000-lb hopper cars.

Development in related technologies accompanied the growth. Use of form-and-fill packaging, as an example, put many powdered or granular food products into polyethylene-coated pouches. Many frozen confections went into similar pouches, replacing the previous paper bag. The introduction of higher density polyethylene resins and polypropylene brought interest in extrusion of barriers against petroleum products, with constructions such as composite cans for lubricating oils, and multiwall bag liners for packaging oiled products.

At present, most polyethylene coating resins cost less than 15¢ per pound and are unlikely to drop further. Market expansion has also dropped off somewhat since 1965, and present growth of the extrusion coating industry depends largely upon growth of already existing markets.

STRUCTURE OF POLYETHYLENE

Polyethylene, polymerized ethylene, can be pictured as a long chain of carbon atoms, each atom linked to a pair of hydrogen atoms, a structure like the following: (Fig. 5-2)
with a general formula of C_nH_{2n+2}, in which (n) will have a value of about 1,600 to 1,700 for most coating grades. The long chains are more or less branched, and the chains themselves vary considerably in length, so the molecular weight of a sample will necessarily be an average. The general formula is the same as that of paraffin wax; polyethylene is, a giant cousin of that material. As the molecular weight decreases, polyethylene becomes more and more like paraffin, with decreased elasticity and flexibility, and a lower and more restricted melting range. Like paraffin, polyethylene in the molten state is quite

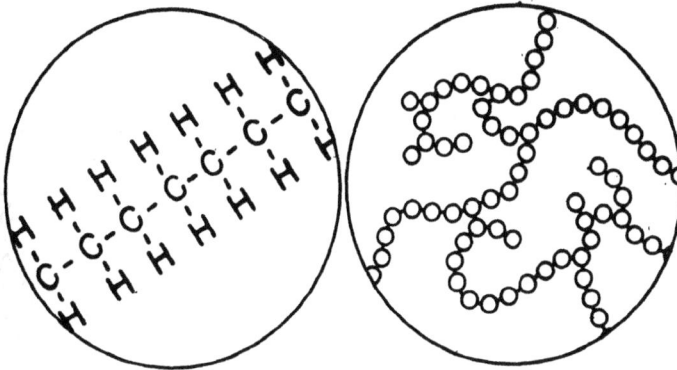

BASIC UNITS   **BRANCHING CHAIN**

Fig. 5-2. Closeup of polyethylene molecular chain, showing (right) complex branching of a typical molecule.

POLYETHYLENE

CRYSTAL

STRUCTURE

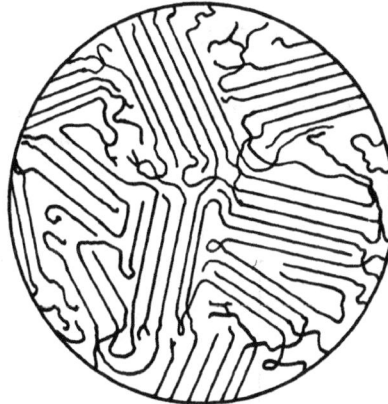

Fig. 5-3. Polymer chains participating in both crystalline and amorphous zones. (Fringed micelle model)

transparent, but becomes hazy upon solidification. The hazy appearance, from light scattering, suggests a two-phase structure within the solid resin. This has been visualized as a partially crystalline and partially amorphous system (a fringed micelle) with each chain participating in a number of crystalline units, and winding randomly among them. (Fig. 5-3.)

Recently[8], a modification of this concept has been proposed, the "folded chain" and "telephone switchboard" models, with the same molecular chain doubling back on itself to form a crystal zone. (Fig. 5-4.)

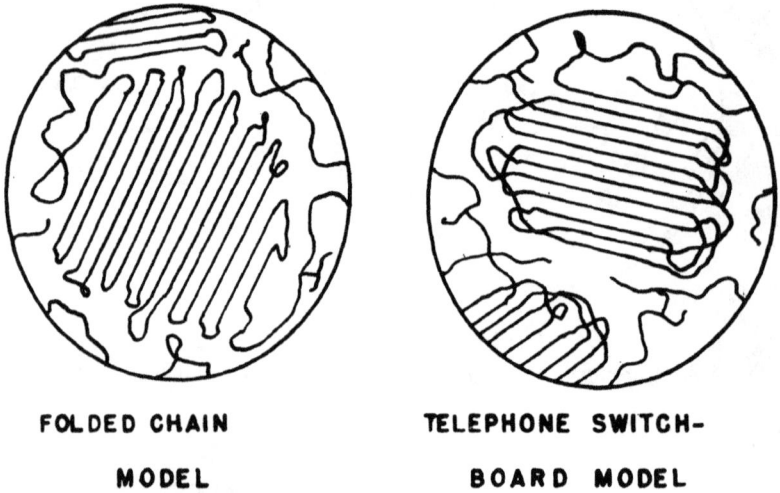

FOLDED CHAIN   **TELEPHONE SWITCH-**
MODEL   **BOARD MODEL**

Fig. 5-4. Two other crystal models that conform to observed properties of polyethylene.

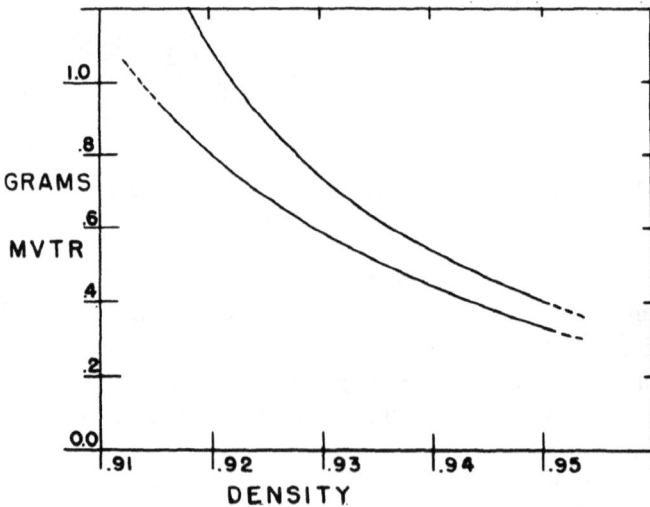

Fig. 5-5. Moisture vapor transmission rate of a one-mil coating of polyethylene (General Foods Method) vs. resin density.

In either case, as the fraction of the molecular units involved in crystal units increases, the density of the plastic also increases, because of better "packing" of the chains, and both rigidity and barrier properties increase accordingly. (Fig. 5-5.)

Manufacturers and buyers of polyethylene resins for coating define the product by two properties: density and melt index. That these are useful descriptive parameters, no one denies; that they provide a complete prediction of resin performance, no one proposes. Since 1958 more or less aggressive search has been made for some other property that would actually predict behavior under extrusion conditions, but the usual method to evaluate a new resin is simply to buy a 1000-lb lot (sometimes as much as 5000-lb for a coating line with a modern appetite) and run it under commercial conditions.[9] What can be learned from density and melt index?

Density

As we have seen, all polyethylene resins have more or less branching in their molecular chains. The branches interfere with orderly alignment of chains into crystal units, and thus increase the average distance between them. Density increases, then, and branching decreases. For this reason, a resin of high density is sometimes called "linear polyethylene", as branching is at a minimum. The amount of branching is controlled during manufacture, and resin of a given density is produced.

Coating resins are offered in the range of densities between 0.914 gm/cc. and about 0.955 gm/cc. Because of imperfect annealing in extrusion, the actual density of a polyethylene coating is slightly lower than of the same resin before coating. Grades offered for sale are divided into three types by density:

Low density	(Type I)	0.910-0.925 gm/cc.
Medium density	(Type II)	0.926-0.940 gm/cc.
High density	(Type III)	0.941-0.965 gm/cc.[10]

Production of Type III polyethylene involves a different process[11] than the other two, so a small gap exists (between about 0.932 and 0.945 gm/cc) in which few coating resins are made.

Resin density, being related to chain branching in the polymer molecule, affects properties of a finished coating, particularly in terms of its barrier capabilities. Low density polyethylene, because of the spaces between chains, and less orderly crystal development, cannot impede the passage of solvents, gases, or water vapor as well as a higher

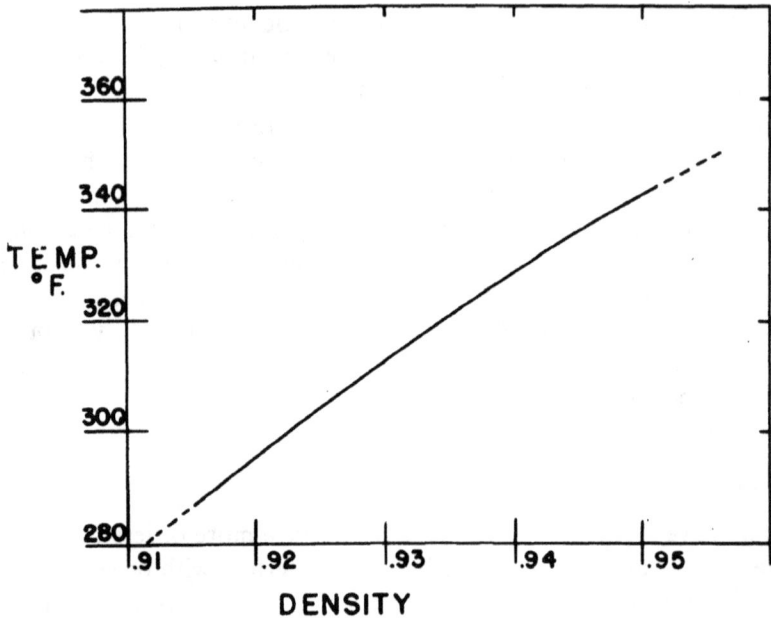

Fig. 5-6. Minimum heatseal temperature vs. resin density — temperature
is that required to make a face/face seal at one second dwell time
at 40 psi between two heated jaws.

density plastic. The temperature at which a resin softens also increases
with density. (Fig. 5-6.) The same orderly crystals that create a better
barrier, however, also account for greater shrinkage upon cooling from
extrusion temperature (about 600°F.) to ambient temperature, resulting
in greater curling of paper coated with high-density plastics.

Reliable heatsealing is also somewhat more difficult with coating of
density of more than 0.950 gm/cc and coatings tend to be slightly more
brittle, although they also resist scratching and abrasion better.

The density of pelleted resin or a separated coating can be tested in
several ways. The standard ASTM test is a water-displacement method,
with a pycnometer.[12] The method is quite accurate, but requires precise technique because of the small difference between the density of
polyethylene and that of water.

In laboratories in which a large number of tests are run, a convenient method utilizes a density gradient tube. A vertical glass column,
about four feet long, is filled with a solvent mix, with the lighter solvent
floating on the heavier. Resin samples dropped into the tube sink

until the density of the surrounding liquid is the same as that of the resin. A number of calibrated glass bubbles floating in the liquid act as reference points for the estimation. A simple method which, can be used in even a small laboratory, is to float the sample material in water in a small beaker and titrate alcohol or acetone into the container until the plastic is at equilibrium with the liquid; then measure the density of the liquid with a Westphahl balance or an accurate hydrometer. Yet another method is to set up a number of large test tubes of alcohol-water mixtures with graduated densities, dropping the samples into the tubes until the right density range is located. (As alcohol is volatile, the tubes should be kept tightly closed when not in use, and be checked with a hydrometer before each test.) All the above procedures require closely controlled temperature for best results. (A TAPPI conditioned test room is recommended.)

Melt Index

Polyethylene, like most thermoplastics, has no definite melting point. A softening range exists, as discussed previously, which varies according to density, but melt index determines, to a cosiderable extent, flow behavior in processing equipment. Melt index is described, according to the ASTM test procedure,[13] as the weight of resin, in grams, that extrudes in ten minutes through a standard orifice (0.0825 inch in diameter and 0.315 inch long) under a fixed pressure (43.5 lb/sq. in) at 190°C. The stated temperature is well above the softening point of all commerical coating polymer resins. The melt index increases as the melt viscosity decreases at the test temperature and is essentially an inverse function of the molecular weight.

Currently, coating resins are available with melt indices between about 3.0 grams and 13.0 grams, although at least one grade with a value of close to 24 has been available for some years. Recently, as coating line speeds have increased, more interest has been shown by converters in resins with higher melt indices.

Melt index is a parameter of more value to the extrusion coater than to many of his customers. (The exception would be those that heat-seal the coated product.) Barrier properties of a coating are largely independent of the melt index, as crystallinity is tied more closely to density. (In heatsealing, however, a high m.i. offers a faster sealing cycle at commercial temperatures. The coated surface also becomes somewhat less slippery with increasing m.i.) During the coating process, consideration of melt index becomes important. As melt index increases, extruder pumping capacity (throughput) increases as well, usually allowing higher coater speeds. This has been especially important in older coating lines, with shorter barrels. Also increasing with melt index, is drawdown, the capacity to form a competent coating at very low basis weights.

Unfortunately, the average molecular weight, as indicated by the melt index tells only an incomplete story of the processing behavior of a resin. One large factor is the pressure specified for the test, as it is far below the 1,000—2,000 lb/sq. in normally prevailing in the interior of a commercial extruder. Another important factor is the behavior of a resin under high-shear conditions, such as between a rotating screw and the wall of an extruder barrel. Two polyethylene resins of similar melt index may have widely differing properties in actual use, principally because of the actual distribution of the molecular weights of the individual chains of the polymer. (Fig. 5-7.) Location of branch chains and perhaps other structural factors also has some effect. Viscosity of molten polyethylene drops rapidly under the shear stress imparted by the rotating screw, and apparently to a greater extent for those resins with a wider molecular weight distribution.[14]

There has been considerable pressure from large polyethylene users* toward the development of some other test that would better predict the processing character of a resin. Many high-pressure viscosity tests

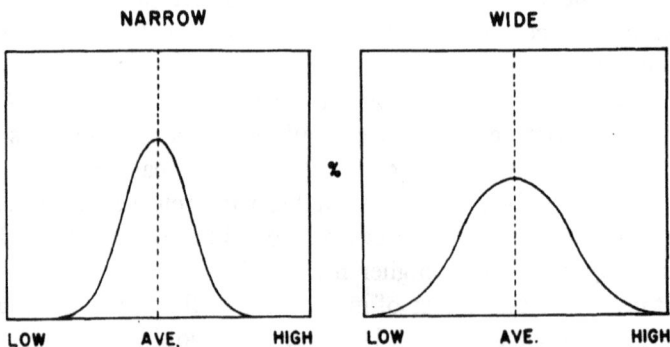

Fig. 5-7. Molecular weight distribution of two resins having the same average molecular weight, with one (right) having a much wider spread between lowest and highest weights in the sample.

---

* Particularly from the Plastics Extrusion Coaters Group of the Specialty Paper and Board Division of the American Paper Institute, Inc.

have been discussed, as well as one involving the elastic memory of a sample in a regular melt indexer, but no new parameters have been accepted, partly because of lack of interest shown by many smaller converters and resin suppliers.[15]

SPECIFIC BARRIER PROPERTIES OF POLYETHYLENE

Most purchasers of polycoated paper have functional needs involving retention of, or protection from, water or aqueous solutions, chemicals, organic solvents, oils, or vapors.

In designing a coating for a set of specific needs, a general view of properties of polyethylene is useful:

Liquid Penetration

Unless high temperatures are anticipated, a pinhole-free coating of polyethylene on paper is an effective barrier to all dry and almost all aqueous materials. Exceptions are very strong oxidizing liquids, such as nitric acid, halogens, and concentrated halogen solutions. Many organic liquids can satisfactorily be contained, as liquids, but practical packaging of many of these is difficult because their vapors pass easily through a polyethylene coating.

Polyethylene is a hydrocarbon of high molecular weight, and its resistance to chemical attack by swelling can generally be considered according to the rule of "Like dissolves like". Polyethylene is most successful as a barrier against highly polar materials, but is increasingly penetrated and finally softened as polarity decreases, in roughly the following order:[16]

1. Alcohols (and glycols)
2. Carboxylic acids
3. Aldehydes and ketones
4. Ester (and glyceryl fats and oils)
5. Ethers
6. Hydrocarbons and halogenated derivatives.

Resistance to attack is heavily dependent upon temperature, something to be considered in package design. Many "snack" foods, for instance, are packaged hot, increasing the penetrating power of their shortening or frying oils.

Moisture Vapor Transmission

A substantial amount of polyethylene-coated paper and paperboard is sold entirely or partly as a barrier against moisture vapor. Although

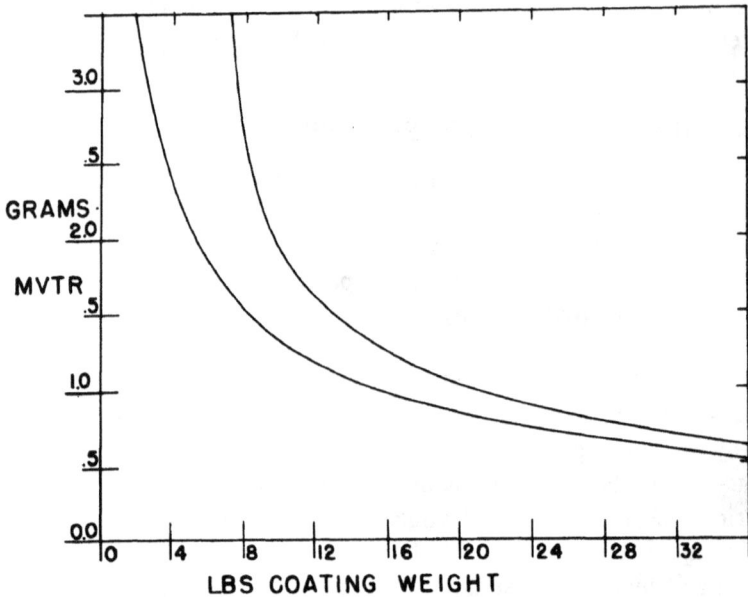

Fig. 5-8. Moisture vapor transmission rate (General Foods Method) vs. coating weight. At lower weights, the effect of substrate roughness and processing variables becomes more pronounced. (14.4 lb/ream is equivalent to a 1 mil coating).

many plastics, including some that are extruded, are superior vapor barriers to even high-density polyethylene, the present low cost of polyethylene gives it a real advantage when thinking of "barrier per dollar". The chart (Fig. 5-8) shows moisture vapor transmission rates* for low-density polyethylene resin, coated on a smooth kraft paper. Values shown are affected, especially at low coating weights, by the smoothness of the substrate paper. Extrusion conditions may also affect moisture transfer, as high resin temperature and excessive nip pressure may increase pinholes and lower the density of the finished coating somewhat. Corona treatment for printing and gluing can also increase the moisture transfer rate, although proper control of current and electrode design can eliminate this problem entirely.

---

* Tested in a General Foods Cabinet maintained at 95% realtive humidity and 100°F, against calcium chloride.[17,18] Several other commonly-used test conditions are described in ASTM E96-63T.

A final.factor in package design is the side of the polyethylene coated sheet exposed to the high humidity. Paper is swelled by moisture, and the expansion stress on the polyethylene coating tends to increase the moisture transmission rate. Thus, a package to protect a product from surrounding high humidity would best be designed with the vapor barrier coating toward the outside, facing the higher humidity. Unfortunately, it is usually impractical (such as in heat-sealed pouches) to face the coating in this way, and a somewhat heavier coating is needed to build a competent package. The high humidity effect decreases when a smoother substrate is used, as fewer cellulose fibers penetrate into the coating.

Gas Barrier

An important reason for interest by converters in coating resins other than polyethylene is its modest barrier effect against atmospheric gases (oxygen and carbon dioxide) and organic vapors.[19] The transmission rate of most gases is relatively high, with the result that polyethylene cannot be used in packaging many organic materials that do not attack

Fig. 5-9. Die and combining nip of a recently installed extrusion coating line. *Courtesy of Thilmany Pulp & Paper Co.*

or penetrate the plastic as liquids, but their vapors pass so readily that
the contents vanish in a short time. This "weakness" of polyethylene
has prompted interest in alternate plastics, even a higher cost, and
development of overcoatings of Saran or similar materials, onto a polymer substrate. (See "Co-extrusion.") One more traditional approach
is to work with polyethylene coated glassine, as glassine paper is a significant gas barrier in itself, especially at high temperatures.[20]

EXTRUSION EQUIPMENT

Extrusion coating machinery is specifically designed to form a relatively thin sheet of molten polyethylene or other thermoplastic and
adhere it to a web of base paper or other sheet material. Polyethylene
or other plastic resin, in pellet or granular form is fed into the extruder
(Fig 5-10) at room temperature, emerging from the extruder as a viscous
melt at approximately 600°F., (for polyethylene; lower for most other
plastics) formed by a slot-shaped die (Fig 5-11) into a sheet at the correct width and thickness for adhesion to a passing web of substrate.

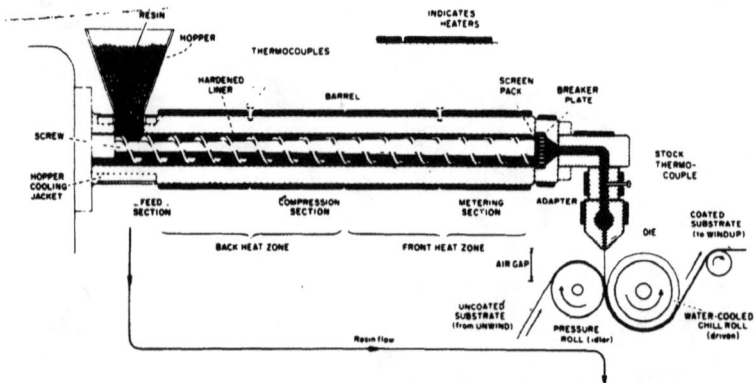

Fig. 5-10. Cross-section of extruder barrel and attached die. *Courtesy of
U.S. Industrial Chemicals Co.*

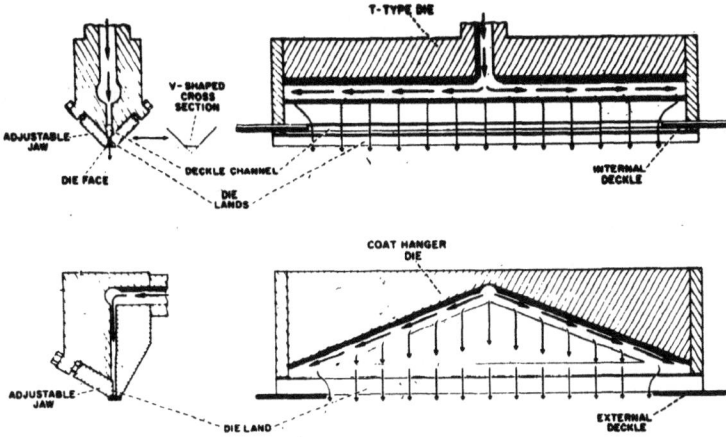

Fig. 5-11. Cross-section of two common die types. The T-type die is used in most commercial installations, partly because of ease of maintenance and adjustment of width. *Courtesy of U.S. Industrial Chemicals Co.*

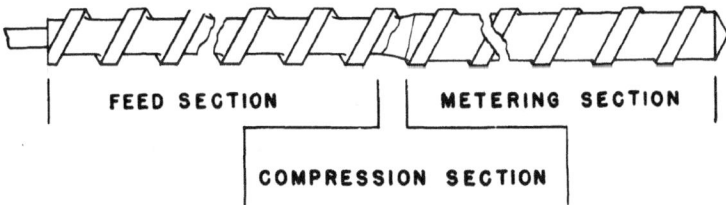

Fig. 5-12. Typical screw for an extrusion coater. Such a screw usually has a length/diameter ratio of 28/1, about equally divided between feed and metering sections, with a short compression section between them.

The extruder itself is a long narrow barrel enclosing a slowly rotating screw. (Fig 5-12.) The entire assembly is covered with electrical heating elements, which assist in melting the plastic resin. (Friction accounts for a large part of the necessary heat, and under some operating conditions, electrical heating is needed only to compensate for radiation losses.) One end of the barrel has a hopper to feed plastic pellets into the unit, and rotation of the screw carries the pellets and the resulting melt toward the other end. The screw itself has a large root and rather shallow flights, with some provision for compressing the mass of plastic. This can be achieved by decreasing pitch of the screw; but most modern screws operate by increasing the root diameter at the points at which compression is desired.

A drive motor at the hopper end rotates the screw, with heavy thrust bearings to counteract axial force. The necessary internal pressure in the melt zones is maintained by the solid pellets entering at the cold end. These tend to melt prematurely when the unit is idling, so most modern screws are bored for at least part of their length for water cooling, to maintain pressure and reduce "bridging" of partially melted pellets in the hopper throat. Along the barrel, a series of thermocouples records and controls temperature.

Proper operating pressure in the exit end is maintained by a "breaker plate," a perforated metal disk set in the orifice. The breaker plate supports one or more layers of metal screening called "screen pack". In older installations, internal pressure (between 1,000 and 2,000 lb/ sq. in and resulting operating performance are altered by varying the number and fineness of the screens used. More modern machines usually have a valve installed in the line, so the screen pack is used principally to catch tramp particles in the resin. (Removal of foreign material is a poor alternative to clean storage and handling.)

A short adapter pipe conducts the plastic melt into the die itself. Heating elements surround this, too, but ideally only for control of radiated heat. The die section forms the resin into its final shape. In most extruders, the entire die assembly is hung onto the end of the adapter section, with the resin entering the die through a manifold in the center. The resin flow splits and runs the length of the heated die, from which it is extruded through a longitudinal slit in the bottom. The slit is partially closed by a pair of adjustable jaws, which are moved for profile control.

The gap is set so that the exit slit measures about 0.016—0.020 inch. In operation, the 0.020-inch sheet of melted resin is brought to its final caliper by passing the substrate under the die faster than the resin is emerging, thus stretching it uniformly by the desired amount. (The relation between the two speeds is called the "drawdown ratio".) Thus, the thickness of a coating may be controlled by varying two factors: the rate of extrusion and the velocity of the passing substrate.

Beneath the die, the resin is pressed to the paper substrate in a combining nip, the polyethylene surface against a water-cooled steel roll (chill roll) which may be polished to a glossy surface, or sand-blasted, to produce a matte finish. The opposite roll of the combining nip is made of a heat-resistant rubber and is movable, to open the nip. The coated web, when cooled by passing around part of the circumference of the chill roll, passes to a windup.

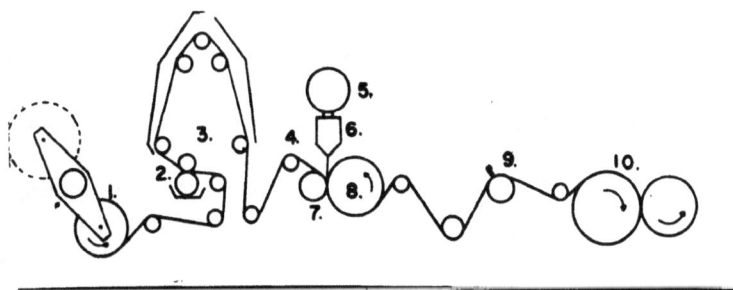

Fig. 5-13. Profile of a typical extrusion coating line.
1. Unwind stand
2. Rotogravure priming station
3. Drying oven for priming station
4. Preheating roll
5. Extruder barrel
6. Die
7. Rubber pressure roll
8. Steel chill roll (water-cooled)
9. Corona treating bar
10. Windup for coated paper

For lamination of two paper webs, an additional unwind stand would be installed to the right of the extruder barrel.

A complete coating line has many other features. (Fig 5-13.) Present operating speeds demand an unwind with a flying splice arrangement for automatic roll changes. Most machines have a rotogravure station ahead of the combining nip, for application of bond-promoting primers. Some machines can use the same station for in-line printing, with the appropriate etched roll and reversing the web through the unit. Almost always, there are one or more heated rolls, or an infrared unit, as warming the paper enhances bonding. A corona treating bar may be installed either ahead of the combining nip, for bond promoting purposes, or behind it, for producing a printable polyethylene coating. If any lamination is anticipated, an additional unwind stand must be present, to supply the second paper web. A few coating lines, mostly in the plants of milk carton producers, are set to coat both sides of the paper web in one pass by means of tandem extrusion dies.

This description has one major variant, an extrusion process that has no combining nip, but instead, adheres the polyethylene film to the base paper by a vacuum through the paper. Some barrier advantage is claimed, noting that a coating applied in this manner covers the irregular paper surface more uniformly, with less chance of moisture-conducting fibers penetrating the coating. Application of the vacuum process is limited to relatively porous base papers and boards, and no variation of finish is possible.

Polyethylene resin may be supplied to the hopper by hand loading from a platform, but more usually from a remote loading area by means of a pneumatic system.[21] Thus, a dry, clean area may be selected for handling the plastic as both dirt and moisture can seriously affect operation.

Polyethylene resin can be purchased in a variety of packages, the choice usually depending upon consumption. For grades used in small quantity, or which must be hand-loaded, 50-lb multiwall bags are convenient, usually shipped as 2,000-lb pallet loads. A convenient and versatile intermediate package is a 2,000-lb corrugated bin, which is unloaded with pneumatic equipment and can be reused in the shop. A limited number of converters purchase their polyethylene in 10,000-lb collapsible rubber balloons, which are handled as hanging units. Resins used in large quantity are delivered in especially designed hopper cars that hold 100,000 to 120,000 lb at a saving in cost, both in the resin itself and in handling, as well as in protection from contamination in delivery. A hopper car can be emptied into a holding silo or into used 1,000-lb boxes until the plastic is needed. A large user may consider direct feeding right from the hopper car, as emptying time might be short enough that no demurrage need be considered.

A diversified converter might be using all of these systems at once, as he would inventory a range of densities, some perhaps in more than one range of melt index. His customers may demand black, white, and possibly other colors, so he will stock these, or concentrates for blending them in his own shop. In addition, more and more coaters are working with polypropylene, vinyl, and polyamide resins, and keep these in inventory as well.

PRIMERS FOR POLYETHYLENE COATING

Polyethylene, having little chemical affinity for paper, has to be coaxed to stick to its substrate in a coating operation. Polyethylene adheres through two mechanisms, one physical and one chemical. When hot polyethylene is pressed to the paper in the combining nip, there is some actual flow of molten resin into the spaces between the paper fibers, developing some adhesion through mingling of the plastic with the paper. This is entirely dependent upon the smoothness of the base sheet, with a rough kraft paper showing more physical bonding than a glassine sheet.

The chemical portion of adhesion is developed principally by the resin in extrusion. The 600°F. temperature at which the polyethylene melt emerges from the die is sufficient to cause some oxygen attack, with a corresponding increase in the chemical activity of the surface. The presence of carbonyl and other polar chemical groups on the surface of extruder polyethylene is thought to cause adhesion to the similarly polar substrate.

Both resin selection and extrusion conditions influence bonding. A higher melt index allows somewhat more resin flow into interstices of the paper substrate and improves mechanical bonding. Preheating of paper also improves bonding, probably by promoting physical flow. Chemical attraction can be increased by raising the melt temperature, so that surface oxidation increases, or by lengthening the exposure time to the air by raising the air gap (distance from die lip to combining nip). Both measures carry some risk that increased oxidation might degrade the resin sufficiently to develop a scorched odor and damage the heatsealing properties of the coating.

Numerous chemical agents have been developed for application to the paper substrate to help generate reliable bonding. These may be applied to the base paper, either in line on the coating line or in a previous operation. Resourceful coaters have also incorporated them into printing inks, tub sizing solutions, or supercalendering dampening water. Paper thus treated can be coated at higher line speeds or a lower operating temperature.

Although many excellent package primers are sold to coaters, some quite effective agents are home-made. Agents in common use are:

1. Organic titanates, which form titanium dioxide by rapid hydrolysis. These have to be kept totally free of water until applied.
2. Dewaxed shellac, either in alcohol or alkaline water solution. Shellac is easy to use and store, and can be incorporated into many inks.
3. Colloidal silica, quite simple to use, but can be abrasive to machine parts if mishandled.
4. Polyethylene imine, either in alcohol or water solution. PEI in alcohol wets foil.

Another priming means, though not strictly chemical, is the application of a high-voltage corona discharge to the paper web just prior to coating. This method is quite effective on most papers, although not suitable for foil, metallized paper, or other conductive substrates.

PRINTING OF POLYETHYLENE COATED PAPERS

Many of the commercially useful properties of polyethylene derive from its low wettability, relatively low response to most solvents, and inertness to many chemical agents. These very properties also prevent reliable adhesion of most printing inks and adhesives. Soon after gegeneral introduction of polyethylene films and coatings active efforts were made to alter somehow a polyethylene surface to make it more receptive to wetting.

One early approach was a chemical system, oxidizing the surface with application of sodium dichromate in sulfuric acid. Another process, particularly suitable for polyethylene film, was chlorination in the presence of light, releasing hydrochloric acid as a by-product. The slightly chlorinated surface was more polar and thus more wettable to inks and glues.

In 1953, two patents were issued,[22,23] describing a heat treatment of film or coated paper under an open gas flame. Although the mechanism was not clear at the time, subsequent work pointed to a surface oxidation, with resultant wettability. A variation to increase the durability of the treatment, in which a thin coating of a wettable material was immediately applied to the flame-treated surface was patented in 1960.[24,25].

Flame treatment has largely been supplanted by electric corona treatment[26,27,28,29], in which the coated web is passed through a high-voltage discharge, the corona being generated between a rubber-covered roll and a bar electrode placed parallel to it and spaced between 1/16 and 1/4 inch away from its surface. Although the effect with respect to acceptance of inks and adhesives is identical to flame treatment, electric treating appears to depend less on oxidation than upon dehydrogenation, so that unsaturated groups in the polymer provide the necessary polarity.

It would be convenient to coat polyethylene film or coated paper with some agent that would adhere to the polymer, and which itself would be a receptive base for subsequent printing or gluing. An active search for such a coating is in progress and may soon be a reality.

Proper treatment is only the first step to an attractive printed sheet. Press room practice must be modified somewhat from plain paper printing. The oil ink processes, offset, letterpress, and silk screen present some major considerations in setting up. One is the barrier aspect of polyethylene. Although a treated polyethylene surface adheres to inks, the surface is completely nonabsorbent and all drying must be from the surface. Consequently, oil inks are usually run with a quite high content of dryer. Ink films tend to be glossy, though in most jobs, this is no disadvantage. (Flatteners may be used, if needed.) Another result of total surfacing of the ink is decreased abrasion resistance. Many ink formulators recommend an overvarnish in every case, to improve the "scratch".

One serious problem, particularly in sheet-fed presses, is sticking and double feeding caused by static electricity. Additionally, static causes dirt and lint to stick aggressively to the coated surface with resultant smudges and voids. Even in humidity-controlled press rooms, an electronic eliminator is a virtual must.[30].

The solvent processes, flexographic and rotogravure, seldom offer any special problems in turning out a good-looking job. Liberal use of copper tinsel and other anti-static means are needed, however, to prevent dangerous sparking.

ADHESIVES AND GLUING

Most commercial adhesives are suitable for gluing treated polyethylene, although the waterproof coating on one or both sides of the glued seam prevents rapid setting. Minimal use of adhesives is desirable. Some latex adhesives are not highly compatible with polyethylene, and testing of a sample is always prudent before purchasing adhesives.

Neither flame or corona treatment is permanent. Wettability decays gradually in the warehouse and can be destroyed easily by careless web handling in converting machinery. Once printed or adhered, however, no further decay occurs. Treated paper should be converted as quickly as practical after delivery, and it is usually worthwhile to recheck the treatment as part of routine set-up procedure in the printing or gluing operation.

Functional testing of the treatment can be quite simple; a dampened strip of gummed paper tape pulls fiber when dry, or a drawdown with ink is able to resist stripping with pressure-sensitive tape. (A test ink is available in a spray can for simple use.) Although no single test has had universal approval, successful operation of a treating line requires some type of quantitative evaluation. For some years one method used a water drop on the treated surface, with tipping of the sample slowly and recording the angle at which the drop started to move, the angle being a function of the contact angle of water and proportional to the wetting tension. Another approach now widely used, is the so-called "Visking test", which directly measures the wetting tension in dynes/cm^2 [31] by a drawdown of a series of fluids (a graduated mixture of formamide and ethylene glycol monethyl ether) with varying attraction to the treated polyethylene. The drawdown breaks into droplets with the fluid mix that fails to wet the polymer surface, thus showing an endpoint at the corresponding treatment level.

EXTRUSION COATING OF PIGMENTED PLASTICS

Of the major applications of thermoplastics, extrusion coating is almost unique in the small amount of colored plastic used. Molded articles are usually colored, electrical insulation and piping are mostly composed of colored resin, and even extruded films are widely bought in black, white, or bright colors, but sales of polyethylene resin for coating are still mostly of clear material. Why? A number of reasons contribute. One is that a colored paper substrate easily substitutes for a colored plastic coating. If paper of the exact shade is not available or practical, an allover tint can be printed onto the sheet, making close matching economical for fairly small shipments. A second reason is that a substantial portion of the coated paper sold is for barrier or heatseal considerations, and the coating is not exposed to the public. For example, the barrier layer of a multiwall bag simply cannot be seen. Additionally, coloring of extrusion resins, aside from black or white, is not easily done. The 600°F temperature of extrusion coating is considerably higher than processing temperatures for molding or film extrusion, and relatively few coloring agents resist the required heat without unpredictable color drifting or outright decomposition. (Some also catalyze decomposition of the polyethylene as well.) Coloring agents that do have sufficient heat stability are often oxides or salts of metals such as cadmium, chromium, or copper, and are thus eliminated from food contact situations. Finally, control problems and relatively small sales have kept resin prices high to converters.

An exception is the case of black or white polyethylene. A brilliant, pure-appearing, white resin can be produced by the addition of titanium dioxide to the clear material, in concentrations usually between ten and twenty percent. Coatings made from such resin can show a uniform white color, even on dark base paper, at between 16 and 20 pounds, 24×36 inches, 500. Actual paper color influences the exact shade at these weights, but the overall effect is a quite satisfactory white. Black polyethylene is colored by addition of a special grade of carbon black, at a concentration of two to three percent.

Use of colored resin may have either decorative or functional ends. Opacity is obvious, especially in the case of black coatings. One common application is in ream wrap for blueprint and other light-sensitive papers. Some photo film packaging also involves black polyethylene. Black polyethylene can also become an economical coating that suggests patent leather in cover papers for jewel boxes and vanity cases. A grade coated with white polyethylene finds a similar use in covering of albums for wedding snapshots. A major use for white polyethylene coated kraft paper has been for a "billboard" effect in packaging of bulk items, such as lumber and steel shapes. A heavy, reinforced packaging paper is coated with white polyethylene and given a large, simple trademark copy which is legible from a flatcar or open truck. Both protection and striking product identification are achieved all the way to the final user and sometimes beyond, as the discarded wrapper often ends up as a tarpaulin at a construction site.

A relatively new market for white polyethylene coatings is in label printing of composite fiber cans, competing with the popular printed aluminum foil label. Up to now, at least, the polyethylene coated label has been largely confined to lubricating oil cans. Another packaging application is in coated liners for bottle jar lids, with the coated paper laminated to a fiberboard backing, from which disks are cut.

Pigmented resin, especially black resin, is an essential factor in coating for outdoor storage or building papers. Polyethylene is severely damaged by sunlight. A one-mil coating does not survive more than a few weeks when exposed to the weather, because of attack by ultraviolet light. Even heavyweight coatings of three mils or more become brittle and waxy in a short time. The use of a black-pigmented resin extends the life of a coating to years instead of weeks, allowing a host of outdoor uses for polyethylene-coated packaging papers. Some of the grades included are:

1. Crate liners for deck and flatcar shipping.
2. Steel shrouding, for both basic shapes and fabricated items.
3. Temporary wall material for construction operations and poultry sheds.
4. Agricultural canopies for stacked straw, hay, and silage.
5. Concrete curing blankets, to retard moisture loss.

Use of other colors in coating has not been really widespread, although a few items have had quite steady sales. One is in nursery wrap, in which a green polyethylene coating on wrapping paper suggests vegetable vigor in plants and shrubs wrapped in it. Some olive drab and khaki resin also finds use in military packaging.

OTHER RESIN ADDITIVES

Use of carbon black is not the only way to achieve protection from ultraviolet light although it is usually most economical. If a clear coating is a requirement, chemical UV inhibitors can be added to the pelleted resin before extrusion, either by a resin supplier or by a converter. Some users feel that the protection is only marginal, especially compared with what can be done with films and molded shapes, in which the lower processing temperature again allows more latitude in selection of additives.

A commonly used additive, which can be bought in a blended resin, or mixed just before extrusion, is a slip agent, a waxy material which "blooms" to the surface and lowers the coefficient of friction. While use of a matte finish, slippery in itself, usually permits satisfactory operation in "form-and-fill" and similar operations, additional slip is sometimes required, or product design does not permit use of a matte finish. A slip additive in the resin lowers the coefficient of friction by as much as a third. (High percentages of slip materials can interfere with heatsealing performance, and should be used with this in mind.)

Antioxidants and chill roll-release agents are also sometimes added to base resins, though these are usually blended by the resin supplier.

In recent years, more and more converters have installed their own blending machinery, buying small quantities of concentrated colors and additives and mixing these with clear resin as needed. Some real inventory savings are possible in this way, and actual cost of resin can sometimes be reduced by larger purchases of base resin. Blending, however, must be thorough and uniform before sending resin to a coating line. The polyethylene extruder is essentially a pump, not a mixer, and incompletely blended resin produces a coating with heavy streaks or even holes.

TESTING OF POLYETHYLENE COATED PAPERS

Routine control testing is no less important to the producer or buyer of polyethylene coated papers than to those that buy or sell any other product. Test information is required by the producer as a guide to material and machine costs and to both producer and user as a guarantee of uniformity and function of the product.

Only a few routine tests are relevant to *all* polyethylene coated products, although there is a lengthy list of possible tests. The usual practice is to choose a set of applicable tests for a given product, and to develop a specification with them. Some of the more common tests are described as follows:

Constituent Weights

Analysis of ream weight on a polyethylene coated sample is no different from testing plain paper or paperboard. To know the coating add-on, however, requires separation of the coating, so that either the coating or the paper can be weighed separately. Separation can be done in various ways, according to convenience and the nature of the product. Polyethylene is soluble in hot chlorinated solvents, with trichloroethylene commonly used in test laboratories. A small machine-shop degreasing unit loaded with trichloroethylene makes a serviceable extractor, and can handle a number of samples at one time. Basis weights of the constituents are taken by difference, with the precaution that the paper plies must be reconditioned to their original moisture, as the extraction desiccates the sheet.

Another procedure especially useful for dense papers and those with low bonding levels (to the coating) is to break the bond by wetting the sample sheet with alcohol or acetone. This method has an advantage in that, by weighing the separated poly film instead of the paper, no reconditioning is necessary other than drying the liquid from the surface.

For double-coated grades, it is sometimes convenient to slip a sheet of silicone coated paper or heavy foil into the extruder nip.[32] The polyethylene applied to this sheet can be easily stripped off and a sized sample weighed directly.

Polyethylene Bonding

What constitutes "good bond" depends on the physical demands of the end use of the product. (This is an area in which over-specifying is very common.) In some cases, severe abuse requires a "fiber-tear" bond, defining a coating that cannot be separated from its paper substrate at all without destroying one of the plies. However, since extruder running speed is often limited by bonding, intelligent definition of bond and production costs are closely related.

Most bonding tests are some kind of tensile test, recording the force required to separate the coating from its substrate. One low-cost device for this purpose is the Keil Tester[33], a simple tensile instrument originally developed by the Dow-Corning Company for evaluating release papers and pressure-sensitive tapes. Other tensile testers for paper are also adaptable for bond testing, but those that read only the maximum value may tell an incomplete story of the product. All tensile-type bond tests are heavily dependent upon paper grain direction, and are usually run crossgrain for best reproductability. One "directionless" instrument is the Perkins-Southwick machine,[34] a device that records the force necessary to separate a coating from its substrate by air pressure, applying the pressure through the porous paper substrate. Combinations having polyethylene on both sides, coated on impermeable substrates, or containing foil cannot, of course, be tested in this way.

A simple test for grades with fiber-tear requirements is to apply a strip of pressure-sensitive tape to the coated surface and then pull off the tape, examining for fibers adhering to the loosened coating. This method has some flexibility, as tape with a more, or less, aggressive adhesive can be selected for individual grades.

Heat-sealing

All extrusion-coated plastics are inherently heat-sealable by being thermoplastic, but the resin itself and processing variables affect performance of even a face-to-face seal. A face-to-back sealing requirement also necessitates some thinking about the substrate paper. Wet-strength paper, plasticized papers, and dense grades usually create special sealing problems. Most converters' test programs include some approximation of the end user's sealing conditions on a laboratory jaw sealer.

Pinholes

Lightweight polyethylene coatings are not continuous, and contain pinpoint holes, according to coating weight, roughness of the substrate, and, to a surprising extent, extrusion conditions. The end user with specific barrier needs must often demand freedom from these pinholes in his purchases, or at least set some limit on their number. Different liquids wet polyethylene more or less aggressively, and exact needs must be known when designing a grade. A pinhole visible in kerosene may not show when tested with alcohol. Common test liquids are alcohol, motor oil, turpentine, and various water/detergent mixtures, all of these dyed for visibility (or permanence) for record keeping of penetration stains. Sometimes, especially in the case of a proprietary mixture, a user may supply a sample of the actual material for use in testing.

Odor

Extrusion of polyolefins at high temperatures may cause a characteristic "burning candle" odor in the final product, something most unwelcome to the packager of food or cosmetic products. Holding a sample of the coated paper in a clean, closed glass jar for fifteen or twenty minutes concentrates the odor sufficiently to allow a go/no go sniff test. Gentle heating (115-120°F) and addition of a few drops of distilled water to the test jar make the test more sensitive. Occasionally, off-odors in the base paper are also detected, although these, of course, cannot be dealt with by changing operating conditions of the extruder.

Treatment for Printing

For some grades, particularly those coated on papers containing groundwood, periodic testing of stored materials is useful. A retest after any rewinding is also indicated, as passage through a machine can occasionally damage the printability.

Physical Testing

Although many users, particularly the United States Government, require their purchases to comply with definite values for burst, tear, tensile, or other physical tests, predicting what will occur when two or more materials of even known strength are adhered together becomes so bewildering that a conservative specification may simply describe the properties of the strongest constituent. A polyethylene coating enhances some test values. Bursting strength increases over that of the base paper alone. Tearing strength also does so, the amount somewhat dependent on bonding. High tear test readings may not indicate a superior product, but are rather a symptom of low bond. Tensile values in the machine direction are seldom affected by a polyethylene coating, although machine creped paper and grades using extensible paper may show some increase. In the cross-grain direction, however, a polyethylene coating usually contributes something. Most grades of paper stretch more under a cross-grain stress, placing some of the load on the coating just prior to failure.

Establishment of test limits should always be the result of fairly complete testing of sufficient samples from different runs, with the tests of plain paper run on exact twin material. This may involve the extraction of the coating from a finished sheet, to assure validity of the base paper.

One aspect of testing cannot qualify as process control testing, as evaluation requires too much time to be relevant as guidance to a coating line in operation. Tests for moisture vapor transmission[18,35] require several days before results are available, and many similar tests, such as those for military specifications, require exposure for a week. In such cases, proper design of the grade must assure compliance, with test results for policing, without hope of active machine control. (Some work has been going on recently with a high-speed test device for moisture vapor testing, but the machine is not yet in general use.)

EXTRUSION RESINS OTHER THAN POLYETHYLENE

Although "polyethylene coating" and "extrusion coating" have been almost synonymous terms, any thermoplastic resin is a possible candidate for the process. Free films of many other thermoplastics have been available for some time, particularly polypropylene, vinyl, polyvinylidene chloride, and nylon, with the experience in film extrusion at least partly applicable to a coating process. Nevertheless, production of such coated papers has not been great. Plastic introductions have had to survive repeated decreases in polyethylene prices and more or less continuous advances in the technology of polyethylene coating. In many coated paper grades, the plastic coating costs less per pound than its paper substrate.

Since about 1961, however, sales of various resins and the technology for their extrusion have been slowly advancing. Polypropylene was one of the first of these,[36] offering high gloss coatings with somewhat superior heat resistance and better ability to survive attack by a greater range of organic materials. (Ethylene glycol and common flavoring oils are commercially important examples). Unfortunately, high density polyethylene resins with some of these same properties became available at the same time, and these had the additional advantage of lower cost and the ability to be processed without modifying converters' machinery. Even at this writing polypropylene and its copolymers, while commercially available as paper coatings, enjoy only those markets in which high density polyethylene lacks some necessary property.

Both nylon and vinyl resins have been extrusion-coated on pilot machinery and, to some extent, commercial coating lines. Either plastic may become important as coatings if resins were better "fitted" for operation on existing extruders, or if markets were to develop that would justify the extensive modification needed for running them. Right now, cover papers are being sold for bookbinding and case items such as jewel boxes with solvent or plastisol vinyl coatings. Such products can certainly be considered for extrusion coating, particularly if converters develop efficient color matching for small lots.

Ethylene-vinyl acetate co-polymers,[37] first developed about 1958 as wax fortifying agents and hot-melt materials. They can also be extruded on paper and other web substrates. These EVA resins have quite competent barrier properties and offer advantages in ultra-fast heat-sealing or where low sealing temperatures must be used.

Another line of products of interest in recent years are the "Ionomers," copolymeric plastics of ethylene and methacrylic acid,[38] containing ionic linkages. The cost of these is presently much higher than that of polyethylene, but Ionomers reliably bond to a variety of difficult substrates and can be drawn to very low (two pounds or less) coating weights. Additionally, gas barrier properties appear quite good, something in which polyethylene is somewhat deficient.

Really substantial sales of these or other plastics will probably first occur in areas in which they compete with other coating processes (organosol, emulsion, solvent or hot melt) rather than competing with polyethylene as an extruded plastic. Some vinyl and similar film applications may offer markets for a coated product, particularly where strength and dimensional stability can be more cheaply secured with a paper base than with additional plastic.

WHERE ARE WE GOING?

Extrusion coating of paper and paperboard is called a "mature industry" in that growth of the industry more or less parallels that of the gross national product. Steady advances in technology have given us improvements in present products and a host of new ones, as well as substantial gains in coating line speed, but few really new attacks on product or machine design limitations have appeared in the last few years. Sales of coated products have continuously expanded, but "big bang" growth spurts (such as coated milk cartons) have not been part of the recent past.

The foregoing paragraph can become false overnight, especially for the individual converter. Even a modest growth item as counted on an industry-wide basis, can sharply stimulate the sales of that converter whose product mix includes that item. There is sure to be further use of composite can grades, as more and more products are considered for canister packaging.

Sure market growth is ahead for polyethylene coated mulch paper as weed and moisture controls in agricultural row crops. A recent engineering development is an extruder assembly that applies a composite coating of two or three different resins,[39] each in a discrete layer and with its own properties, "Co-extrusion".

Designers of folding cartons have not given up thoughts of a reclosable barrier carton without the loose inner liner of waxed or similar paper. A milk carton-type package can certainly house a wider variety of liquids, perhaps prepared batters for baking, and particulate solids, such as garden fertilizers or bread crumbs. With improvements in corona treatment or an anchor coating for a polyethylene surface, there would be increased use for a smooth, pinhole-free, coated grade as a substrate for other functional coatings, such as silicone or a variety of decorative coats. Although extrusion resins other than polyethylene have not often lived up to expectations, these materials can be made to do many jobs that polyethylene cannot and merit further work.

Past growth has largely been a product of individual effort by resin supplier, machinery manufacturer, papermaker, or converter. Future growth will more likely be the reward of cooperative ventures with all these agencies as participants. Unlike *Homo Sapiens*, a mature industry can undergo rejuvenation, as the right raw materials, using the right machinery, become the right product, for uses not yet even considered.

REFERENCES

1. Frados, J. and Evans, C.J.W., *Paper Trade Journal* 152, no. 12: 48 (Mar. 18, 1968)

2. Mosher, O.D. *TAPPI 5*, no. 206: 9-A (May, 1962)
3. Billmeyer, F.W. *Textbook of Polymer Chemistry* New York: Interscience Publishers, Inc., p. 30 (1957)
4. Ibid, p. 328
5. Kresser, T.O.J. *Polyethylene* New York: Reinhold Publishing Corp., pp 8-9 (1957)
6. Renfrew, A., and Morgan, P. ed. *Polythene* pp 4-5 Interscience Publishers, Inc., New York: (1957)
7. Ibid, p. 6
8. Oppenlander, G.C. "Structure and Properties of Crystalline Polymers" *Science* 159, no. 3821: 1311-1319 (Mar. 22, 1968)
9. Dowd, L.E. and Cameron, R.A. *Extrusion Coating for Profit.* Paper, Film and Foil Converter 42, no. 11: 94-97 (Nov., 1968)
10. ASTM Standard D1248-58T
11. Renfrew, Op. Cit. p. 9
12. ASTM Standard D792-50
13. ASTM Standard D1238-62T
14. *Selections from Coating Memos*, Gulf Oil Company Plastics Division, Kansas City, Mo.
15. Anon. *PE Coaters Push for More Resin Supplier Parameters.* Modern Converter 12, no. 9: 1 (May 5, 1968)
16. Pinsky, J. *Modern Plastics* 34, no. 145 (1957)
17. Symington, F.S. and Burroughs, R.F. "The GFMVT (Southwick) Test for Moisture Vapor Transmission", *Fibre Containers* 28, no. 4: 107-109 (April, 1943)
18. TAPPI Routine Control Method RC-318
19. Stannett, V. et al., TAPPI Monograph no. 23, *Permeability of Plastic Films and Coated Papers to Gases and Vapors.* p. 96 Technical Association of the Pulp and Paper Industry, New York; (1962)
20. Ibid, p 45
21. Whitlock, D. *Trends in Bulk Resin Handling*, Paper Film and Foil Converter 42, no. 11: 64-66 (November, 1968)
22. Kreidel, U.S. Patent no. 2,632,921
23. Krichever, U.S. Patent no. 2,648,097
24. Rice, et al., U.S. Patent 2,955,970
25. Rice, et al., U.S. Patent 3,076,720
26. Traver, U.S. Patent 2,910,723
27. Traver, U.S. Patent 3,018,189
28. Traver, U.S. Patent 3,113,208
29. Mosher, et al. Canadian Patent no. 687,395
30. Young, L., *Litho Printer* 2, no. 3: 11-12 (March, 1959)
31. Promulgated by the Plastics Extrusion Coaters Group of the Specialty Paper and Board Division of the American Paper Institute, Inc. At this writing, the procedure is being considered as a TAPPI Routine Control Method.
32. TAPPI Routine Control Method RC-323
33. TAPPI Routine Control Method RC-283
34. Guillotte, J.E., and Mac Dermott, C.P. *Bond Tester for Coatings*, Modern Packaging (December, 1956)

35. TAPPI Test Method T448 m-49
36. Kresser, T.O.J. *Polypropylene* p. 128 Reinhold Publishing Corp., New York (1960)
37. Perino, D.A., *New Resins—New Functions—New Markets*, Paper Film and Foil Converters 42, no. 8: 76 (August, 1968)
38. Ibid
39. Anon., *Extrusion Casting of Composite Films*, Modern Plastics 45, no. 8: 130-134 (April, 1968)

BIBLIOGRAPHY

Evans, J.C.W. Ed. *Trends in Paper and Paperboard Converting*, Lockwood Trade Journal Co., New York (1965)

Newgarden, A. Ed. *Know Your Packaging Materials* American Management Association, New York (1958)

Frados, J. Ed. *Modern Plastics Encyclopedia* McGraw-Hill, Inc., New York Revised and published annually

chapter 6

The Use of The Computer
In New Product Development*

MICHAEL C. SACHER

INTRODUCTION

For many years it has been customary for new product development groups to make judgments and set priorities between projects by utilizing such parameters as (1) availability of manufacturing equipment vs. new capital investment, (2) apparent readiness of the marketplace to accept one product vs. another, (3) the ease of overcoming technical problems in one project vs. those present in others, and (4) the pressure on the new product group generated by the needs of the marketplace as transmitted by the marketing and sales organization. In addition it has always been possible to make a choice between projects by utilizing a weighted number evaluation system or similar problem-solving exercise, although it has been difficult to make clear-cut decisions—black or white in nature—where the "what if" factor was involved. Where manufacturing costs, size of market, and equipment availability were all known factors, such a black or white decision could be made with some reasonable accuracy. If the sums of the weighted numbers were close together, however, and instead of a sharp black or white interface one achieved only a variety of gray colorations, an earlier day decision could only be made based on experience and intuition. This is one of the great values offered by the computer! The ability to evalu-

---

* All numbers in this chapter are disguised, and for illustration purposes only. In fact, the input data shown were not run through the computer, so the output will not check accurately against them. The teaching of the concept, however, is not invalidated by using disguised data.

187

ate the "what ifs" and to modify and change the value of a large number of business parameters simultaneously and in quick sequence permits the sophisticated new product development group to decide accurately and wisely among projects that may appear under cursory examination to offer little choice. If the proper programming has been developed for the computer, the project manager now has a tool to make decisions. To confirm these decisions over a period of time the input data can be revised and updated as a result of the availability of additional technical or market information.

Because the reader might be misled by the chapter title, a broader definition might be "The Use of Mathematical or Financial Analysis in New Product Development." This distinction is made because most people use their computers today mainly as marketing and business control tools, through such functions as order handling and billing procedures, payroll and employee benefits computations, inventory control, and distribution analysis. Only the more sophisticated use it in "optimizing" a set of variables into the most efficient of all possible combinations and permutations, to arrive at good marketing or other business decisions.

Several companies have gone further yet and developed computer programs designed to compare one new product project or business venture with another, or even several others, utilizing the common denominator of return on investment as the basis for comparison.

Such programs are very flexible in that they can be used to test for the "sensitivity" of profitability to a change in value of any one of several variables. To put it simply, one can "ask" the computer "what if" questions. For example: What if manufacturing costs are 10 per cent higher than they were estimated to be? What if sales volume is only half as large as projected? What if a 20 per cent higher selling price can be obtained than estimated?—20 per cent lower? Implied in all of these questions, of course, is the question of what happens to the profitability of the project if these unforeseen events occur.

This ability to test the profitability of the project on the computer to determine its sensitive areas—i.e., if is it more sensitive to volume changes than to price changes, or vice versa—is of great value in alerting the project manager to potential trouble spots.

GATHERING DATA

People familiar with computers, use the well-known word, GIGO. This word is an abbreviated form of the expression "garbage in, garbage out" and is a humorous way of indicating that the output of the computer is only as good as its input.

Gathering good data can be a difficult task. Good market research (See Chapter 2) is needed to determine the size of the market for a new product, its growth trend, the share of that market a new product can command, and the potential price range within which the product must compete. Sizes, colors, put-ups, channels of distribution, and advertising and promotional programs desired by potential customers, all of which affect costs, may be required knowledge prior to making a computer run.

Internally, data must be gathered on manufacturing cost estimates (usually at several volume levels), marketing expense, general and administrative charges, freight policies, and costs.

Finally, for a new product, educated "guesstimates" must be made for future years regarding the length of the product life cycle, expected changes in sales volume, possible deterioration in selling prices as, or if, competition develops similar products, the effect of increasing or decreasing volume on manufacturing costs, and the necessity and effect on profitability of supportive capital investments.

PRINTOUT—INPUT DATA

The printout of the input data from the computer allows the manager to "audit" the input for potential errors.

TABLE 6-1
One Proposed Computer Printout Format
Input Data
Profit And Loss—Rate of Return Analysis

Project_____ Manager_____ Date_____ Run No._____

Case_____ Job_____

Cost of Manufacturing Adjusted By _____ Dollars Per _____.
Selling Price Multiplied By _____ In Each Year Of Study.
Unit Volumes Multiplied By _____ In Each Year Of Study.

Year   Selling Price   Cost of Mfg.   Volume Capital   Dev. Exp.   Other Exp.

SCF FKC RA FW DIR GA AP TR CAPCR PE FRT CAPEX

Depr. Deduct Option—If 1, On—O
DDB Option—If 1, On—O

TABLE 6-2
One Proposed Computer Printout Format
Output Data
Profit And Loss—Rate of Return Analysis

Year	Net Sales	Cost of Sales	Devl. Exp.	Depr.	Capital Wk. Cap.	Nt. Prft.	ROS	ROA	ROR

To fill out the forms with suitable numbers, one can assume that the data gathering has been completed, (See Table 6-3), and is as follows: The project is WIDGETS, the manager is Joe Smith, the date is January 31, 1970, and this is the first run. In this case, after careful study, Plant A has been chosen to manufacture the product, but in sheets instead of rolls. Arbitrarily sheets are assigned the number "2" to distinguish them from other products. Since this is the first run, the cost of manufacturing is adjusted by zero dollars per ton, selling price multiplied by one, and unit volume multiplied by one, in each year of study. In later (cases) runs, when testing for sensitivity the printout will so indicate in these spaces; e.g., unit volumes would be multiplied by 0.5 in each year of study, if one wished to ask "What if the volume is only half of that projected?" In the present case, however, the form begins to look like this:

PROFIT AND LOSS—RATE OF RETURN ANALYSIS
Project *WIDGETS* Manager *JOE SMITH* Date *1/31/70* Run No. *1*
Case *PLANT A* Job *2 SHEETS*

Cost of Manufacturing Adjusted By 0 Dollars Per *Ton*. Selling Price Multiplied By 1 In Each Year Of Study. Unit Volumes Multiplied By 1 In Each Year Of Study. Now a project life of five years has been estimated, and the data indicate a selling price of $500 per ton; cost of manufacturing is $300 per ton; sales volume will be 1,000 tons in Year 1, increasing to 5,000 tons in Year 5, with a capital investment in new equipment of $50,000 required in Year 2, development expense

TABLE 6-3
One Proposed Computer Printout Format
Input Data
Profit and Loss—Rate of Return Analysis

Project WIDGETS

Case PLANT A

Manager JOE SMITH

Job 2 SHEETS

Date 1/31/70

Run No. 1

Cost of Manufacturing Adjusted By 0 Dollars Per TON .
Selling Price Multiplied By 1 In Each Year Of Study.
Unit Volumes Multiplied By 1 In Each Year Of Study.

Year	Selling Price	Cost of Mfg.	Volume	Capital	Dev. Exp.	Other Exp.
1	500	300	1,000	0	25,000	0
2	500	300	2,000	50,000	25,000	0
3	500	300	3,000	0	25,000	0
4	500	300	4,000	0	25,000	0
5	500	300	5,000	0	25,000	0

SCF	FKC	RA	FW	DIR	GA	AP	TR	CAPCR	PE	FRT	CAPEX
1.0000	1.0000	0.0100	0.0200	0.3000	0.0400	0.0500	0.0500	0.0700	0.0100	20.0000	200.0000

Depr. Deduct Option—If 1, On—O
DDB Option—If 1, On—O

of $25,000 per year for each year, no "other expense" incurred, and
$1,000 of working capital. Thus, the printout further develops:

Year	Selling Price	Cost of Mfg.	Volume	Capital	Dev. Exp.	Other Exp.
1	500	300	1,000	0	25,000	0
2	500	300	2,000	50,000	25,000	0
3	500	300	3,000	0	25,000	0
4	500	300	4,000	0	25,000	0
5	500	300	5,000	0	25,000	0

What the reader sees above is "raw input data," and it is specified for
each year of project life. Once the length of project life is determined
(in this case 5 years) there are six pieces of the "raw input data," expressed as follows:

ITEM	EXPRESSED
Volume	In whatever units chosen; e.g., tons
Selling Price	Dollars per unit
Manufacturing Cost*	Dollars per unit
New Capital	Dollars invested per year
Development Expense	Dollars incurred each year
Other Expense	Dollars incurred each year

*May or may not include depreciation.

The program is designed to give the option of calculating (a) straight
line, (b) double declining balance, or (c) no depreciation.

The remaining symbols on this input sheet are "adjustments to raw
data," and are defined as follows:

SCF—Scale factor—factor used to adjust volume schedule without
changing raw data—SCF of 0.5 halves all volumes shown.

FKC—The percent of selling price returned to the corporation after
adjustment for discounts. (This can be another type of scale factor.)

RA—Returns and allowances.

FW—Freight and warehousing—internal.

DIR—Direct selling expense.

GA—General and administrative overhead.

AP—Advertising and Promotion expense.

TR—Income tax rate.

CAPCR—Capital credit—investment credit against taxes allowed in
first year of capital life.

PE—Plant and equipment—maintenance.

Working capital—Dollars required to support inventories and accounts

TABLE 6-4
One Proposed Coding Form

EACH LINE OF DATA REPRESENTS 1 TAB CARD

CARD TYPE 1

PROJECT TITLE	MANAGER	DATE	RUN NUMBER	VOL. UNIT NAME	PAGE
WIDGETS	JOE SMITH	1/31/70	1	TON	1 01

CARD TYPE 2

NO PRODUCTS ENTERED OR BE ENTERED	TOTAL NO. PRODUCTS	PROJECT LIFE	BASIC RUN DESCRIPTION	PAGE
1	1	1	Plant A JOB 2 SHEETS	1 02

CARD TYPE 3

PRODUCT NO.	YEAR	SELLING PRICE ($/UNIT)	MFG. COST ($/UNIT)	SALES VOLUME (UNITS)	CAPITAL (NEW $)	DEVEL. EXPENSE ($)	OTHER EXPENSE ($)	PAGE NO.
	01	500.00	300.00	1000	0	25000	0	03
	02	500.00	300.00	2000	50000	25000	0	04
	03	500.00	300.00	3000	0	25000	0	05
	04	500.00	300.00	4000	0	25000	0	06
	05	500.00	300.00	5000	0	25000	0	07
	06							08
	07							09
	08							10
	09							11
	10							12
	11							13
	12							14
	13							15
	14							16
	15							17
	16							18
	17							19
	18							20
	19							21
	20							22

TABLE 6-5
Second Proposed Coding Form

EACH LINE OF DATA REPRESENTS 1 TAB CARD

CARD TYPE 4

#	ADDITIONAL DESCRIPTION OF THIS VARIATION	MFG. COST ADJUST. ($/UNIT)	CALC. DEPR. Y-1 N-0	USE DDB Y-1 N-0	DEPREC. OVER ?YRS.	PAGE NO.
1	STANDARD RUN NO CHANGE					0.1
2	DECREASE SELLING PRICE 10 PERCENT PER TON					0.3
3	INCREASE MFG. COST $10 PER TON	+10				0.5
4						0.7
5						0.9
6						11
7						13
8						15
9						17
10						19
11						21
12						23

CARD TYPE 5

#	VOLUME SCALE FACTOR	PRICE SCALE FACTOR	RET. & ALLOW. (% GR. SALE)	FRT. & WHSING. (% NT SALES)	DIR. SELL EXPENSE (% NET SALES)	GEN. & ADM. OVERHEAD (% NET SALES)	ADV. & PROMOTION (% NET SALES)	INCOME TAX RATE	INVEST CREDIT RATE	BUDGET PLANT & EQUIP.	1ST YEAR WORKING CAP ($)	FRT. ALLOWED ($ UNIT)	EXISTING CAPITAL ($ UNIT)	EXIST CAP AS LUMP SUM ($)	LAST CD INDIC. LC=1	PAGE NO.
1	1	1	.0100	.0200	.0300	.0400	.0500	.500	.07	.01	1000	20.00	200.00			0.02
2	1	.09	.0100	.0200	.0300	.0400	.0500	.500	.07	.01	1000	20.00	200.00			0.04
3	1	1.	.0100	.0200	.0300	.0400	.0500	.500	.07	.01	1000	20.00	200.00			06
4																08
5																10
6																12
7																14
8																16
9																18
10																20
11																22
12																24

receivable—specified in first year of project life—second and later years calculated as a fixed percentage of the difference between sales this year and that of the previous year.

FRT—Customer freight—deducted from sales on dollars per unit basis—when freight is allowed.

CAPEX—Investment in existing capital on dollars per unit basis (may also be expressed as OCAP—investment in existing capital expressed as a lump sum.)

Next observe the following factors:

Scale factor is *1* because on the first run we do not wish to alter volume; FKC is *1* (one hundred per cent) because no discounts will be given; returns and allowances, freight and warehousing, direct selling expense, general and administrative charges, and advertising and promotion are respectively *1* through *5* per cent; tax rate is *50* per cent; capital credit is *7* per cent; plant and equipment is *1* per cent; first year working capital $1,000; freight is $20 per ton; and CAPEX is $200 per ton*; the final printout format of input data looks like this.

Following are examples of input cards completed with the above information, and also "asking for" two alternative runs—one with a 10 per cent lower selling price—the other with a $10 per ton added manufacturing cost. (Tables 6-4 and 6-5.)

PRINTOUT—OUTPUT DATA

All this preliminary work, of course, is necessary to obtain the output. It is this output, or end product, of the calculating power of the computer utilizing the input data, that is really the end objective, so let us examine it. (See Table 6-6.)

CONCLUSIONS

Utilization of a carefully written and detailed computer program thus results in a high-speed, formalized system for comparing the projected profitability estimates of various new product proposals with one another. It is also a technique (simulation) whereby the project manager can "play the game" without having to pay what can be the major consequences of a bad financial or even a capital commitment.

---

* In this example, CAPEX will not be depreciated. In actual practice, it can be, thus giving the project manager the option of treating existing capital as new capital for purposes of analysis. When this is done, however, depreciation expense must be removed from manufacturing cost, to avoid "double counting." Cash flow will also be affected.

TABLE 6-6
Computer Printout—Output

Year	Net Sales	Cost Of Sales	Dev. Exp.	Capital	Depr.	Wk. Cap.	Net Profit	ROS	ROA	ROR
1	500,000	300,000	25,000	200,000	0	0	250,000	10	20	40
2	1,000,000	600,000	25,000	400,000	10,000	10,000	120,000	12	25	45
3	1,000,000	900,000	25,000	600,000	11,000	15,000	160,000	13	26	48
4	1,000,000	1,200,000	25,000	800,000	12,000	20,000	220,000	13.5	28	50
5	1,000,000	1,500,000	25,000	1,000,000	13,000	25,000	280,000	14	30	52
TOT	4,500,000	4,500,000	125,000	3,000,000	46,000	70,000	830,000			

The program is also valuable in that it takes into consideration the change in the value of money over a period of time. Thus the dollars of cash flow generated early in the program are treated as having a higher value than those returned later in the project life. This is known as the "discounted cash flow" method of computing rate of return, and may be viewed as an interest rate earned by an investment through cash flows generated by the investment. The project generating the highest rate of interest or return, would be considered the most preferred, other things being equal. Rate of return, ROR, for each year is based on discontinuing the project in that year and recovering the remaining nondepreciated asset value plus all the working capital.

Finally, one of the great advantages of this technique is that once the data have been correctly gathered almost anyone can use it, because it is not difficult to fill out the forms. A knowledge of computers and programming is not required. As stated earlier, the most difficult part of using this technique involves the need to arrive at accurate input estimates, for the data coming out are only as good as the data going in. The effort expended in gathering such data can be well justified and they should be gathered in any case to make an intelligent new product analysis. In total, the program should be viewed as being dynamic, not static. It can easily and quickly be updated as the available data become increasingly more sophisticated and accurate through the continuous development of the project. One should not wait, however, to use it until the project is well under way, because it is also a valuable tool early in the process as a screening device.

chapter 7

Pulp-Molding - - - Three-Dimensional Paper Products

Paper is made by filtering fibers out of a dilute slurry onto a flat screen. Pulp moldings are formed in a similar process, but the screen and the felt that it produces are in the shape of the article being made.

Pulp moldings may be classified as soft, hard, and fully molded. Soft moldings are those that are finished by little more than oven drying. These are fruit packs, planting pots, flower pots, minnow buckets, bottle jackets, egg trays, molded packing inserts, and pie plates. Hard moldings are those that are dried under pressure or dried and remolded under low pressure. These are tropical helmets, spherical world globes, textile cones, luggage shells, TV cabinet backs, car glove boxes, instrument panels, and artillery shell casings. Fully molded parts are high

TABLE 7-1
Pulp Molded Articles Arranged by Density

Class	Preform Treatment	Density (Grams/cc.)	Article
Soft	Oven-dried	0.2–0.5	Shaped packing Egg cartons Fruit packs Decoy ducks
Hard	Dried in heated dies	0.5–1.2	Tropical helmets Luggage shells Car glove box TV back
Fully Molded	Impregnated with resin and remolded at high Pressure	1.2–1.35	Restaurant trays Bowls, dishes Housings

198

in resin content and have been remolded at high pressure. Restaurant trays, dishes and housings are made by this process. A classification of the three groups by density is given in Table 7-1.

HISTORY OF PULP MOLDING

The origin of pulp molding is obscure. It seems, however, to have been developed in the last 150 years. According to *Papier Zeitung*[1] a home industry for the fabrication of plied and glued paperboard snuff-boxes began in the Saar Region of Germany around 1700. A manufacturing plant using the process started in 1839. Large objects, such as trays, hassocks, and stands, were made there. Some time later—the article does not give a date—true pulp molding was begun. Pails, bowls, and tubs were felted in the characteristic unitary seamless construction. In this new process, a pump sucked the water of the pulp slurry through a screened mold of the shape of the desired article. The fibers were caught on the outside, while water went to the inside. Layer after layer of fiber was sucked on until the desired thickness of the shaped mat was obtained. The preform was pressed while on the screen, to smooth it and to reduce the water content. It was then dried in an oven, impregnated with varnish, cured in an oven, polished and finally lacquered.

Verpackungs Rundshau[2] stated in an article that pulp molding originated around 1888. The date is too recent, however, as the following patent shows.

Seamless Boxes

A United States patent was issued in 1856 to French and Frost[3] who disclosed apparatus by which seamless boxes could be made directly from fiber slurries. The machine is shown in Figure 7-1. The fiber slurries were held in vats *B* and *C*.

The screened mold was fastened inside the lower portion of the cylinder shown, at *A*. Vacuum to cause fiber accretion was generated by raising the piston. After the preform had been made, the cylinder was withdrawn from the slurry and moved over the receiving form at the right. Then a reverse stroke of the piston disengaged and deposited the preform. In this method the female felting die had to be set rather high in the cylinder, above a surrounding annular trough that caught the water separated from the fiber. This water was drained from the cylinder before the felting cycle was repeated.

The patent definitely did not mark the birth of the art. French and Frost expressly disclaimed invention of seamless boxes from pulp with

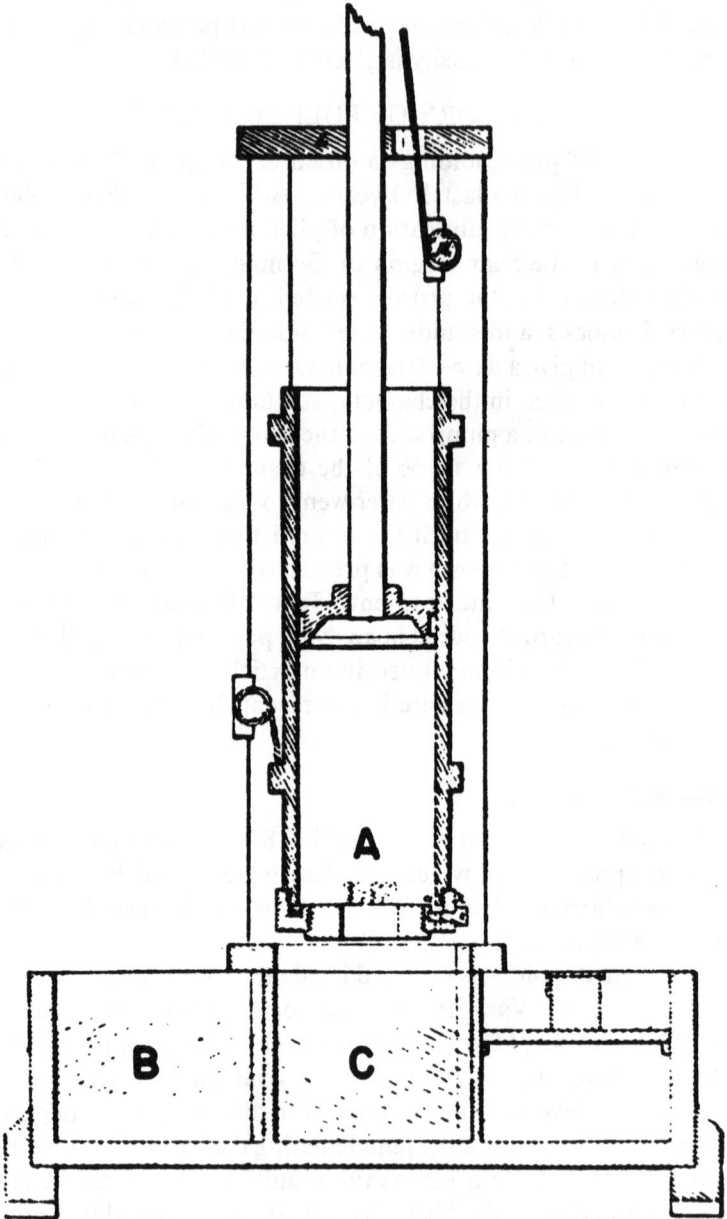

Fig. 7-1. U.S. Patent No. 15,228, July, 1856 — Paper-molding machine,
A. French and C. Frost.

a perforated mold by means of a vacuum. What they did claim was the receiving form, the trough around the die to hold the separated water, and the disengagement of the felted preform by forcing air upon it.

Laminated Pulp Molded Structures

Two vats of pulp are shown in Figure 7-1, and French and Frost pointed out the possibilities of drawing fibers from each to form a laminated box. The lower end of the cylinder was first plunged into the finer pulp and a thin layer of it was deposited on the inside of the mold. This layer formed a coating for the outside of the article. The cylinder was next plunged into the coarser pulp and a thick layer of fiber was deposited, forming the body of the article.

Vacuum Piped to Felting Mold

Knight[4] in 1866 disclosed another form of pulp-molding apparatus. Again both vacuum and air pressure were supplied by the movement of a piston. In Knight's equipment, however, connection to the felting mold was by flexible hose, which allowed the awkward arrangement used by French and Frost to be simplified. The unencumbered mold or molds could now be lowered into the slurry on a movable platform, or set into a cylinder revolving partially immersed in the felting tank.

Tapered Boxes, Edges, Integral Cover and Drying

Knight recognized that pulp molding was adapted to making fruit boxes. These could be made of coarse, harsh fibers and be cheap enough for one use. He recommended that they should have a slight taper so that when empty they would nest. To improve appearance and eliminate the warping that occurs in the drying oven, the boxes should be dried on heated metal forms. The edges of the felted box could be trimmed to shape, or they could be turned down and glued for extra strength. The felting molds also could be united so that a box and lid with a fiber hinge attaching them would be formed. The lid mold could be designed to give a lid with a meeting or lapping joint. To increase translucency at any point, or to make a weak place for breaking, the felting screen could be blocked off at that particular point. Knight's patent is worth reading today for the insight he displayed in the possibilities of the pulp-molding process.

Wheeler and Jerome[5] in 1867 described a screen which fitted over the perforated molding die during felting. This screen supported the felt as it was removed from the mold and aided in carrying the felt to a

pressing mold.   The screen meshes made a drainage system that carried
the expressed water away and finally served as a support for the article
in the drying oven.   In 1872 Kendall and Trested provided a vacuum
for the felting operation by means of a steam jet.   Hotchkiss[7] in 1885
described a method in which a pulp-molded barrel was pressed in
grooved and perforated tools while heated air or superheated steam
was passed through it for drying.   Starr[8] in 1890 felted a pail shape and
gave it a pressure squeeze to compact it and to remove water while the
felt was under vacuum.   Carmichael[9] in 1893 pressed the wet felt of a
box with plates driven by separate hydraulic cylinders. (Pulp catches
between the plate in this system and makes an unattractive part.)   Between 1870 and 1900 there was a spate of patents issued covering machines for felting preform and pressing them to remove water before
oven drying.

Automatic Pulp Molding Machinery

The development of automatic felting equipment by Keyes[10] in 1903
was a significant step forward for the industry.   A drawing from the

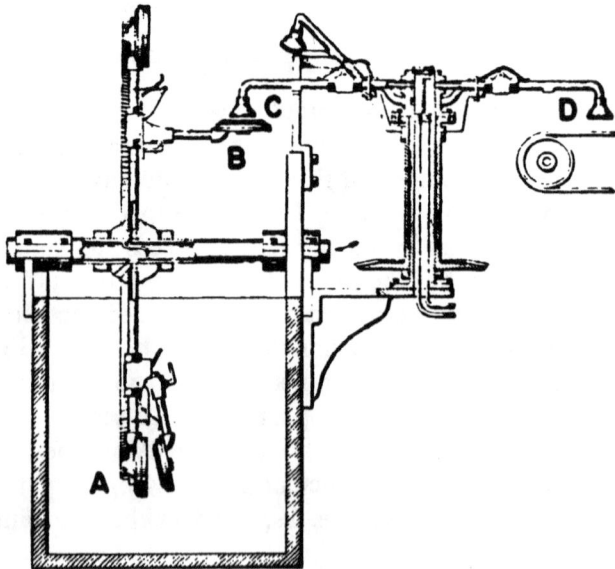

Fig. 7-2.   U.S. Patent 740,023, September 29, 1903 — Apparatus for
making pulp articles, M.L. Keyes.

patent is shown in Figure 7-2. Although only two molds show here, a number radiated from the horizontal axis. The screened felting molds rotated so that while one *A* was at the bottom of the felting tank, another was delivering its felt, *B* to *C* to *D*, to the oven for drying. In the perfected machine, as the mold left the felting slurry, the hinged mating mold shown in the diagram was pressed against it to dewater the preform. Vacuum was applied to this mating mold and the felting die was given a puff of compressed air. This transferred the preform from the felting mold to the mating mold. A similar transfer was used to take the preform to the belt. Many patents have since been issued on the design of automatic pulp molding machines.

Drying in Heated Dies

The pulp-molded preform carries the detail and screen mark on one side as imparted by the screened molding die. On the other side the fiber pattern is ordinarily rough and unfinished. Some of this roughness or bark can be removed by wet pressing. A still greater improvement in appearance is obtained if the preform is dried under moderate pressure in heated, vented dies. Manson[11] and Chaplin[12] developed the die drying process around 1928. Pellegrino and co-workers[13] successfully automated the die-drying process in 1964.

Pressure Felting

An interesting pulp-molding technique has developed in which the fiber slurry is injected under pressure into a screened and drained mold cavity. The part formed is dewatered or even completely dried by the

Fig. 7-3. Pulp-molded decoy duck. *Courtesy of General Fibre Co.*

passage of heated air while it is still in the mold. The split mold is then separated and the molded part is taken out. After oven drying it is usually hardened by a varnish or lacquer dip. The process can produce intricate shapes such as screw lid containers and decoy animals. A pressure-felted decoy duck is shown in Figure 7-3. The mold screen pattern is, however, printed on the surface. Pressure felting developed between 1910 and 1930. U.S. Patent 1,284,937 serves as an introduction to the technique.

STATISTICS OF PULP MOLDING

The statistics of the pulp-molding industry are not clearly worked out. The United States Department of Commerce gives a standard industrial classification (SIC) 2646 of "Pressed and Molded Pulp Goods" as follows: Fiber conduits, cups, dishes, egg case filler flats, egg cartons, papier mâché, plates, pressed fiber products from wood pulp, pulp products, spoons, and utensils from pressed and molded pulp. Value of shipments in this classification was supplied by the Hartford, Connecticut, Office of the Department of Commerce:

Pressed and Molded Pulp Goods (SIC) 2646
(In Millions of Dollars)

1958	–	80	1962	–	97
1959	–	90	1963	–	112
1960	–	87	1964	–	118
1961	–	89	1965	–	140
			1966	–	146

Unfortunately fiber conduit and papier mâché are included in these figures. Fiber conduit is usually made by wrapping a wet paper web on a mandril, and papier mâché is molded from heavy pulp or laid up on a form by means of glued strips of paper; neither operation is pulp molding.

The "Annual Survey of Manufacturers" (from the U.S. Bureau of the Census, Annual Survey of Manufacture for 1966, Value of Shipments by Classes of Products, M66 [AS]-2, U.S. Government Printing Office, Washington, D.C.) separates the data to a certain extent, as follows:

Value of Shipments of Product Classes
for 1966, 1965 and 1964
(In Millions of Dollars)

SIC Code		1966	1965	1964
2646	Pressed and molded pulp goods	147	139	118

| 26461 | Bituminous fiber pipe and sewer drainage and conduit, including fittings of molded pulp | 25 | 22 | 24 |
| 26462 | Other pressed and molded pulp goods | 122 | 117 | 94 |

TECHNOLOGY OF PULP MOLDING

Preparation of Fiber Slurries

Raw Materials

Pulp molding uses fibers familiar to the paper industry: northern and southern krafts, ground wood, rayon, cotton linters, sulfite, over-issue newsprint, overrun waxed paper, and glass. Wet scrap from the process is easily redispersed and used. If the scrap part has been resinated, it requires hammermilling before it can be reslurried.

The preform as felted is at 70 to 80% water content, yet it is immediately exposed to some fairly rough treatment. It is blown off the die by a blast of compressed air; it is placed in an oven where it must be reasonably self-supporting while drying; or, it is exposed to a shearing action as drying dies close on it. Initial wet strength (green strength) of the preform is therefore quite important. Long-fibered kraft imparts good initial wet strength. A synthetic rayon fiber made in form of a tiny flat ribbon, American Viscose RD 101, is even more effective. Steamed, defiberized pulps containing the original lignin from the tree can be cut and beaten until they have fairly good initial wet strength, but then they are too slow draining to use in pulp molding.

Bulk is a valuable property in a preform that is to be die-dried or remolded, because it helps to fill out the die cavity. Bulk is obtained by using lightly beaten fibers, and by introducing coarse 'fibers' such as shredded wood. Coarse fibers lie across layers that occur in a felt and tend to bond them together. Coarse fibers also help to create the porosity that is required of the felt in the die-drying process. When the hot die closes on the wet preform, a blast of steam must pass through it to leave via the drainage system of the mating die. A nonporous felt is often blown apart by this surge of steam.

Beating

Stocks for pulp molding are dispersed in water and the fibers then cut to the desired length. Heavy beating or "hydration" is almost never

used because thick mats are felted in the process and a slow stock seals
off before providing the required thickness. Accordingly, beating
equipment can be rather simple. A breaker or tub equipped with pro-
pellers opens or disperses stock quite well, particularly if water at 140°F.
is used. Consistencies are carried high in the breaker. At 5% consistency churning stock defiberizes itself. The opened stock while in
the breaker may be circulated through Bolton-Emerson, Claflin, or
other cutting type equipment to give the desired fiber length. The
papermaker's beater or Hollander is still quite useful in that it disperses
and cuts stock in one operation. It can easily be controlled; however,
it is slow in comparison with the newer equipment.

Beater Sizing and Resination

Many pulp moldings are hardened for service by being dipped in
solvent varnishes and oven-cured. The solvent makes this a dangerous
operation. When possible it is safer to put the resin in the stock before
felting it. Sizing and resination of stock are done in different degrees.
The first step is the addition of wax emulsion and rosin soap, setting
these with alum as in a normal paper operation. Waxed paper is often
used as a substitute for wax emulsion. The next step is the addition of
one of the many wet-strength resins; this might be a colloidal melamine
formaldehyde, which has been given a cationic charge, so that it will
plate out on the anionically charged paper fiber. Wax, rosin, and
cationic melamine resin satisfactorily waterproof flower vases or minnow
buckets.

When it is desired to bind the felt together to give it more strength,
polyvinyl acetate emulsion is added to the stock. Around 6%, depending on the freeness, can be added to fiber that is to be die-dried.
Too heavy an addition results in felts that seal over in the drying and
puff up with steam pressure when the dies are opened.

The proper polyvinyl acetate emulsion, one that has not been over-
stabilized, deposits on fiber in the wet at 140°F. in the presence of 2%
cationic melamine wet-strength resin. This deposition was named the
Bardac process by American Cyanamid in 1945. More recently the
polyacrylamides have been used to increase the bonding of the resin
particle to the fiber.

In preparing slurries that are to be felted into preforms and then
dried and remolded in solid molding dies, one-stage phenolic resins are
used. These may be added as emulsions, or as powdered resin. High
additions are often made, with more resin than fiber present. The resin

powders are simply cofelted with the fibers. Although quite a bit of the powder may pass the screen, reuse of the white water makes certain that it will ultimately all be caught in the felt.

Williams[14] synthesized cationic wet strength resins, which are effective Bardac type precipitants. They have another more interesting action. When 2% resin is added to the stock slurry, based on fiber, and the slurry heated to 170°F., the wetting character of the fiber is altered. The fiber is still water wetting and will remain dispersed in water. Areas along it which are oil or solvent wetting, have been developed. Such fibers then directly accept solvent solutions or liquid oily materials. Solvents used should be water-immiscible. The oil should be poured directly into the mixing stock. Pigmented polyester resins, for example, have been added to fiber slurries, and cured in the wet to give fade-resistant colors in the preform. Almost no color is lost in the white-water. In a similar manner, slurries have been treated with drying varnishes. Thick blocks of pulp that carried 5% varnish solids have been felted. When these are formed and dried, an amazingly strong, low-density material is obtained. · As a varnish-treated pulp is sticky, the screened felting mold should be given a light lining of regular fiber, before the treated pulp is felted, as a parting agent. Stickiness leaves the pulp on drying and oven curing.

Stock Distribution.

Because the long fibered, free-draining stocks used in pulp molding tend to plug pipelines, one should use 3-or 4-inch diameter pipe. Valving should be with ball valves. These should never be throttled back, but set either full on or off. Elbows should be avoided in the lines, and easy bends employed. Some systems have air or water piped directly into the lines to aid in clearing them when necessary. A line should empty when not in use. The pumps are often of the wide-clearance centrifugal type. When such a pump is shut off, a vertical line will drain back through it and thus remain free of fiber.

The dispersed fiber slurry from the beater or breaker is pumped to a stock chest and diluted to marks, or by metered water addition, to the desired consistency, usually 1%. Agitation is provided in the stock chest to keep the fibers from settling. Off center, slow sweeps are used; being off center, they work somewhat counter to the main tank vortex and keep the fiber well dispersed. The sweeps are often supplemented with compressed air, which may be fed into perforated pipe laid on the bottom of the tank. Stock tanks are used in pairs so that one

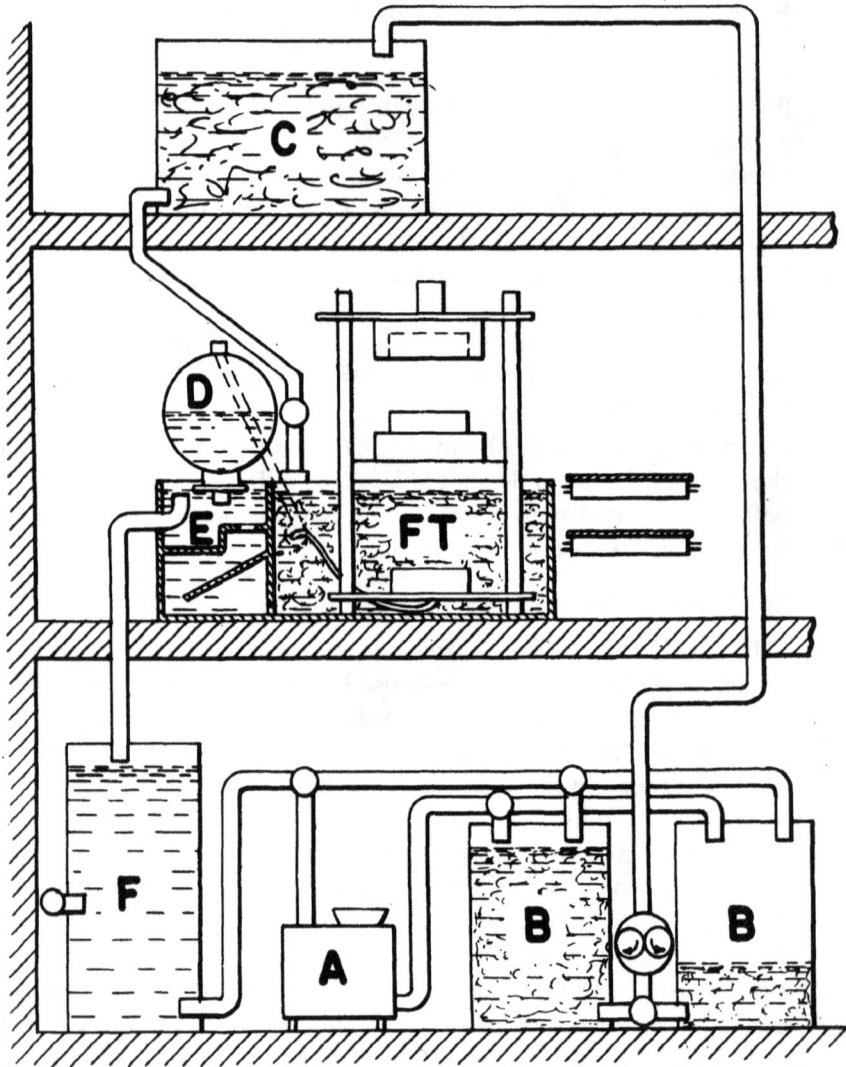

Fig. 7-4. U.S. Patent 3,147,180, September 1, 1964 — Automatic molding apparatus for forming pulp articles, F. G. Pellegrino, *et al.*

handles the requirements of the felting tanks while the other is being refilled.

To maintain the felting tank at the desired consistency when a preform is felted and removed, enough stock slurry must be added to replace it exactly. Handford[15] did this by a timed slurry addition from a circulating stock line. Circulating stock lines, however, are not well regarded because the constant pumping involved breaks up glass fiber and even continues the beating of the cellulose fiber. A better method is to pump slurry into a container to a specified level by probe control, with the contents then dropped into the felting tank each cycle.

A complete stock addition system is shown in Figure 7-4, taken from Pellegrino[13]. Stock is prepared in breaker A, and stored in stock chests B, which supply overhead tank C on probe demand. Metered amounts of slurry are dropped from C into the felting tank FT once each cycle to make up for the fiber being taken with the felted preform. The whitewater from the felting operation is pulled by vacuum to tank D. When the felted part is blown from the felting mold and the vacuum broken, the whitewater drops through the valve shown at the base of tank D into tank E. From here it returns to the felting tank through the baffle system. Excess whitewater overflows through the pipe shown into basement storage F. The level of this overflow pipe determines the level of the felting tank.

Felting Tank.

The felting tank FT in Figure 7-4 has a platform to carry the screened molding die vertically in and out of the slurry. Such platforms are usually moved by a hydraulic piston. Some force is often required as a large brass felting die weighs hundreds of pounds. Keyes used the rotary device shown in Figure 7-2 to move the molds in and out of the felting tank. He was working with small pie plates, whereas Pellegrino's equipment can handle luggage shells or chairs. Mayne[16] held the felting die stationary and raised the level of the felting tank to cover it when felting. Many other methods appear in the patent literature.

Felting is carried out with a vacuum of 10 to 20 inches mercury. This is ordinarily developed by a rotary water-seal vacuum pump, although piston vacuum pumps are not unknown. Whitewater is not allowed to go through the pump, but is separated out as in tank D in Figure 7-4. When fast stocks are being felted in large molds, the system is called on to handle large volumes of water. Piping then must be of large diameter and should lead directly down through the bottom of the tank into a capacious receiving tank. The white water can be continuously pumped out of such tanks against vacuum to keep them free. The conduit may go through the piston rod of the platform drive or through a flexible hose.

The felting platform is also supplied with compressed air, which is used to break the preform away from the molding die. The air is used once, while the vacuum is still on, to clear excess water out of the valving and other parts of the felting platform. Otherwise, this water rewets the preform as it is being blown off and weakens it.

The felting tank is provided with agitation to keep the fibers in suspension; it must be adjustable. Low agitation permits long-fibered stocks to clump and settle. High agitation washes the forming felt off the screened mold. Drilled pipe is laid in the bottom of the felting tank for air agitation. Some form of impeller provides mechanical agitation. Baffles are set in the tank as desired to protect the felt.

The level of the felting tank is maintained by a float or electric probe, which governs the return of the whitewater. The ingenious system shown in Figure 7-4 has been commented on; this holds the felting tank level constant without any attention or special equipment.

The weight of the felted preform is critical in pulp molding, particularly if the preform is to be remolded in dies of fixed aperture. Only one weight will satisfy such a set of tools. Too heavy a preform will tear, or stand the dies off. Too light a preform will not be pressed when the dies close. Preform weight is controlled by freeness of the stock, consistency, vacuum, and time of felting. To increase the weight of the preform it may be desired to raise the consistency of the felting tank. A direct stock line is provided at the felting tank for that purpose. To make a lighter preform, the automatic stock additions are restrained and a number of preforms are felted out of the tank to reduce consistency.

A regular weighing of felts is carried out to monitor the felting operation. Because some felts require an hour to dry in the oven, it is more practical to use wet weights rather than dry weights. Under a given set of conditions there is a fixed ratio between wet and dry weights of the felt.

The preform usually contains 75% water as it is felted, varying with stock and felting technique. Water content can be reduced to 70% by throwing a thin rubber blanket over the felt while the vacuum is still on. This amounts to a presqueeze at about 12 lb/sq in. Shaped latex forms also have been used to give the felt an automatic blanketing. The preform emerging from the slurry enters the shaped rubber for the pre-squeeze. When the vacuum shuts off, the rubber form is released and a counterweight carries it out of the way.

Although the initial wet strength of the preform is governed by the stock formulation, the amount of water in the preform also has a great effect. The wet strength of a stock can be readily measured by pulling a small pad apart. As shown in the following table, a 3.5-gram (dry basis) felted pad three inches in diameter was blanketed to the desired water content. Large spring paper clips were set 1/4 inch apart across the center of the pad. It was next hung from one clip and the other was loaded at a regular rate by pouring sand into a container hung from it. The following wet breaks were obtained from a kraft-shredded wood furnish.

Percentage of Water	Grams to Break Pad
80	150
75	350
60	1500
50	1800
60	2000

These figures show that the drier the felt is made at the tank, the more it resists the rigors of the process.

The interior of the tank should be kept as uncluttered as possible to avoid places where fiber can deposit. Tanks made out of ceramic tile or stainless steel are relatively easy to keep clean. When white articles are to be produced, the whole piping system at the tank should be made of stainless steel. Bits of pulp caught in the whitewater sump reappear at the most inopportune moments and bear a testimony of rust. Finally, the felting tank must have a drain and a water supply to use in cleaning.

The Screened Mold or Felting Die

The screened mold must be strongly built to hold its shape in service. It is loaded heavily when the rubber blanket is thrown over the preform. The best molds are made from brass castings that have been machined to shape, drilled, and then screened. Screen may be attached by solder or by spot welds. Sometimes the brass casting is faced with a perforated metal spinning, which is then screened. Screen is particularly useful with a fine pulp as used in the manufacture of diaphragms for radio speakers. Coarse fibers felt well directly on perforated metal. No matter how clean the mold is kept, it must be torn down periodically and rescreened, because fiber works between screen and casting and can hardly be dislodged.

The felting mold is usually heavy enough to hold itself in place on the platform. A rubber gasket is placed under it to give a good seal to the vacuum and air blow off, which are placed in a small well in the platform. If the felt is to be picked up by a transfer, however, it must be locked securely in place. In addition it should carry guide rods to bring the transfer in precisely.

Interesting felting molds were proposed by Louisot[17], Low[18] and Sheffield[19] in which metal plates were shaped in profile and then bolted together with spacers to provide drainage aperture. These structures would support the perforated metal spinning and screen. For short runs molds may be made of porous sintered metal, or of graded grit particles, resin-bonded.

As far as possible the felting mold should be of one metal. With the large areas involved and salt conductivity in the water, electric currents can be generated that speed the felter on its way to being plugged, and later to disintegration.

Felting molds should be regularly cleaned as part of the felting cycle. This may be accomplished by a needle water spray sweeping across the surface, or by backwashing. Backwashing can be carried out by allowing the mold to submerge without vacuum. When it is filled with water from slurry, the blow-off air is briefly valved on to sweep the water out. The vacuum is next applied and felting commences.

Box[20] made the felting dies the anode in a low-voltage circuit and found that a current of 450 milliamperes kept a 25-inch luggage mold clean in a stock containing polyvinyl acetate and cationic melamine resin as sizing.

Comments on the Felting Process

Felting is a filtration process and the preform is the filter cake. Screened molds usually have a great deal of vertical surface. As the mat cake collects on these surfaces, gravity and tank agitation tend to strip it away. The interwoven character of the mat, its resistance to flow, and the flow being forced through it tend to hold it in place. As the mold is brought out of the felting tank, vacuum drops off as air is sucked into the die. This is a critical moment; many a preform has sloughed away and fallen before it is wholly out of the tank. Here, it may be mentioned, is the great difference between paper making and pulp molding. Paper, being made horizontally is well supported as long as it is on the wire.   Pulp molding, which creates a heavy wet mat on vertical surfaces, requires a great deal of vacuum to be successful. This accounts for the late development of pulp molding; it could not be invented until good vacuum sources were available.

The best preform formation is obtained by felting from dilute slurries, as in paper making.   Dilute slurries under 0.1% consistency, are used in making diaphragms for radio speakers.   However, in most pulp molding the felting tank is run at high consistency, at 0.7% or even above.   The thick pulp molded mat does not require the best structure. Clumped fiber felts quite satisfactorily.

The felt is diffuse as it is formed under the slurry surface.   Its density increases as it is brought out of the tank, creating the rough bark exterior at the same time.   During felting, long fibers are quickly immobilized against the face of the die or the growing mat.   Fine fibers, however, readily pass through the diffuse structure of long fibers.   The fines pack toward the felting mold surface, and the preform is, therefore, more dense on the screen side and open on the bark side.   This graded density of the pulp molded preform is particularly useful in the manufacture of depth type filters.   The open side of the filter mat accepts dirt, while the screen side holds it and prevents it from going downstream.

Having one side of the preform open makes it difficult to paint that side.   A latex base paint usually will seal the surface in one coat when suitably pigmented.

Variation In Thickness

Portions of sharp radii on a male felter tend to be thinner than the rest of the preform.   In a female felter they are thicker and of a poor, "bridged" formation.   This extra thickness, however, is useful in reinforcing the curved portions of a luggage shell.   Male felts are thicker at the base because this part of the die is the first in the tank and the last out.   Mayne[21] attempted to correct this situation in an interesting manner when he lowered an air-tight bell around the raised felting die. The lower edges of the bell were immersed in slurry.   When vacuum was applied to the die, stock was sucked up in the bell to cover the felter and a preform was accreted.   Raising the base of the bell above tank level caused the slurry to drop out of the bell so that felting immediately stopped.

The female felter tends to give felts thick at the top.   The felter comes out of the tank filled with slurry.   This last stock felts largely on the bottom of the felt, which thickens the top of the part. The effect is most prominent with high-consistency slurries.

Elimination of Bark

The shrink of the male felt as it is brought out of the felting tank gives the felt its characteristic wavy bark appearance. Bark can be eliminated by pulling a thin rubber blanket over the felt just before it breaks water coming out of the felting tank. The shrink age of the blanket guides the collapse of the felt to give a smooth surface. Long fibers increase bark; short fibers minimize it.

The Die-Drying Process

Drying the wet preform under pressure in heated and vented dies with vacuum applied increases the felt density and strength. It also improves the appearance of the article by smoothing down the rough bark surface. Die-drying has been used to produce dishes, trays, radio speaker diaphragms, TV cabinet backs, furniture, molded luggage, and

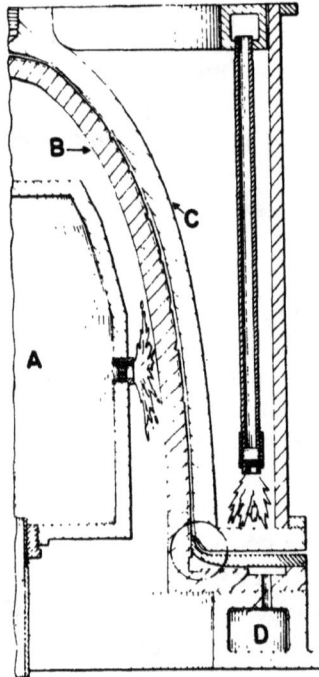

Fig. 7-5. U.S. Patent 3,250,839, May 10, 1966 — Process for making fibrous articles, P.L. DeLuca.

car glove compartments. So much heat is required to dry in dies because of the steady cycling of wet felts that electrical heat is uneconomical. Premix gas heat is usually chosen. A drying die is shown in Figure 7-5. *A* is a manifold supplying premixed gas to the burner tip. *B* is the drained male die and *C* is the solid female die. The male die *B* is grooved for drainage and covered with a fitting spinning of perforated metal. The grooves are brought to a vacuum ring *D* as shown. The burner tips must be placed with some care or the die will scorch the part. It is standard technique to run drying dies with the heat full on and to take the pressed preform out as soon as it is dry and just before it would be damaged by the heat. A slower cycle with temperature-controlled dies naturally makes a better product.

As Figure 7-5 shows, drying dies are complicated; they are hard to make and therefore expensive. There is a good deal of art in getting the correct number of drainage grooves, fitting the perforated metal spinning and placing the burner tips. The die often must be revised on the job until its performance is at the peak. Once checked out, it will produce for years, however.

Die drying is a pressing rather then a true molding operation. The preform does not flow. It elongates as the dies close on it, and it tears. Die and preform must fit closely for pressing to be effective. Horizontal sections of the preform receive press pressure as the dies close. Vertical sections are pressed by wedge pressure. As wedge pressure develops the tendency is to pull the preform down and elongate it. With a preform that is too heavy, wedge pressure can become very high. If the preform does not tear, the press will be stalled so that the horizontal sections of the preform are not pressed. With a preform that is too light, no wedge pressure develops and vertical sections are unpressed. One solution to the problem has been to press very lightly, at no more than 15 lb/sq. in and subsequently harden the part by dipping it in varnish. Another, and better solution, is to presqueeze the felt until it is strong enough to stand up to "high" pressure and die-dry it at 100 lb/sq, in.

In die-drying, a 1/4-inch thick felt is pressed down to 1/8 inch; for the female die to descend on such a bulky felt without dragging or tearing it, the leading edge of the female die is flared out. Internal detail in the die should also be rounded for much the same reason.

The felt is usually placed on the male, drained die and held there by vacuum as the dies are closed. This gives it maximum support. There is a tremendous evolution of steam as the dies close. A great deal of this must pass through the felt into the drainage system. The com-

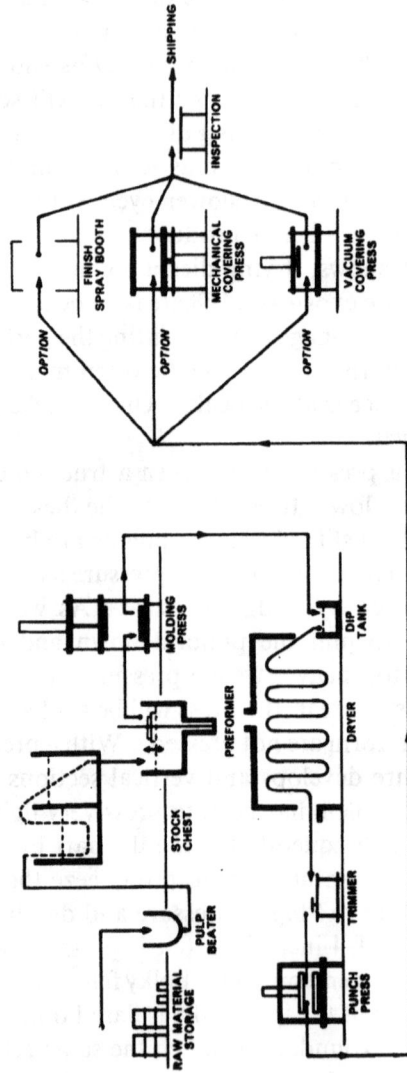

Fig. 7-6A.   Hawley Products — Hitco, Flow sheet for die dry process.

position of the stock is critical. It must have good initial wet strength to stand up to the closing of the dies, yet it also must be quite porous, to pass the rush of steam. Coarse shredded wood is added to give the required porosity.

The elongation of the felt in the drying die means that its lower margin never ends up exactly in the same place. This makes it very difficult to mold the edge. If the dies are designed to mold an edge, the felt either wrinkles badly at that point or does not reach it completely all the way around. Shearing dies have been tried, but they are impossible to keep true when using gas heat. Wet fiber, also, is not easy to shear. The difficulty has led to saw-trimming or die-cutting the edge. The dies are simply left open at the bottom and a ragged edge is produced. The part is then placed on a form for sawing. The form is rotated and the saw follows the form as a guide.

The Hawley Products-Hitco use of the die-dry process is shown in their diagram in Figure 7-6 A.

Muller Process

The Muller process[22] operates with a presqueezed felt, which eliminates many of the problems of die drying. The stock is pressure-felted in a female mold. Some air is then blown through the tools and felt to clear excess water. The felt is then transferred to a perforated male form and, while on this, is given a female bag presqueeze at six atmospheres. Pressure is next reduced to the point that the rubber bag just holds the preform, and the male form is removed. In the next step the female bag is expanded and the preform is dropped to a receiving form which may be the male drying die. The felt is at around 50% water content after this treatment. Die drying is carried out at 100 lb/sq. in. The relatively dry felt does not elongate, so the edge can be molded; there is a slight flash which must be removed. Now the part is hard enough for use as die-dried; it does not require impregnation. It is, however, dipped in plasticized polyvinyl chloride latex to seal it. The Muller process is used to make instrument panels, glove boxes, and gear box covers for automobiles.

Finishing the Die Dried Part

The die-dried part is often given a paint finish. Sealing undercoats have been made from polyvinyl acetate emulsions. Finish coats are usually stipples, wrinkles, or multicolors. Many parts are also covered with cloth or cloth-backed, embossed vinyl film. The cloth is coated with high-molecular-weight polyvinyl acetate emulsion, dried, and cut to size. It is then placed on a draw ring in a "hat press." The male is a fitted rubber bag supported internally by a casting. As the press closes, the pulp molded die-dry part is driven into the female die taking the cloth, under tension, with it. The male rubber bag is then inflated, causing the cloth to be heat-sealed to the part. The use of the high molecular weight polyvinyl acetate gives heat seals strong at the temperature at which they are made, bonds that hold the stretched fabric in place. Excess cloth can be rotary-blade-trimmed on a rotating fixture similar to that used for sawing the die-dried edge.

Dry Molded Preforms

The preform to be molded is felted with the desired percentage of one-stage phenolic resin and dried in a low-temperature oven that does not overadvance the resin. It is then rehumidified and molded.

The molding dies are electrically heated. They do not evaporate a great deal of water, so they have nothing like the energy requirement of the die-dried process. One molding die can produce at the rate of two or three drying dies. The solid metal dry molding die is much less complicated and less expensive than the die for drying. It requires no perforated metal spinnings, no vacuum system, no grooving, no premix gas system, and because of better temperature control, the die can be designed to clip the edge of the part, if pressure is available.

The relationship between resin content, tensile strength, and water absorption in a dry molding is shown in the following data and chart, Figure 7-6 B, from Haslanger and Mosher[23].

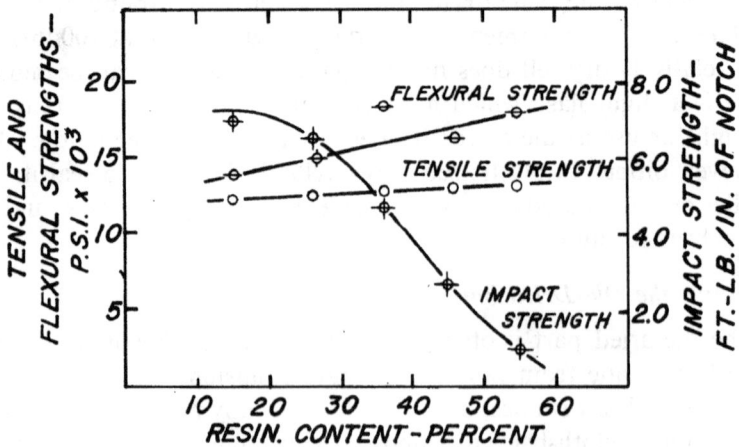

Fig. 7-6B. Dry molding. *Courtesy of Haslanger and Mosher.*

TABLE 7-2

(From Haslanger and Mosher[23])

The Influence of Resin Content on the Physical Properties of Kraft Pulp-Phenolic Resin Preforms Molded at 800 lb/sq in and 320° F.

Resin content, %	Tensile strength, lb/sq in	Flexural strength, lb/sq in	Impact strength, notched Izod, ft-lb/in of notch	Water absorption, 24 hr, %
55	13,400±1130	18,200±1050	1.09±0.18	0.42
45	13,500±1170	16,400±710	2.72±0.16	0.64
35	12,600±1140	18,600±1365	4.80±0.57	0.82
25	12,600±330	15,200±575	6.64±0.78	24.3
15	12,200±1030	14,000±740	6.96±0.64	70.6

Comparison of the Physical Properties of Pulp-Resin Preforms Molded at 880 lb/sq in and Standard High Impact Molding Compounds Molded at 3000 lb/sq in

Material	Resin content, %	Tensile strength, lb/sq in	Flexural strength, lb/sq in	Impact strength, notched Izod, ft-lb/in of notch	Water absorption, 24 hr, %
Kraft pulp-resin preform	55	13,400±1130	18,200±1050	1.09±0.00	0.42
60% pulp, 40% macerated fabric-resin preform	55	13,500±689	16,100±1300	2.34±0.03	0.60
Macerated fabric-resin preform	55	9500±705	16,200±490	3.00±0.08	0.31
Macerated fabric filled molding compound	50	5500±855	10,200±550	3.50±0.13	1.67
Cotton cord filled molding compound	50	4800±625	12,400±1050	8.00±0.18	1.91

In some respects the dry molding process is like the die-dry process and suffers from its difficulties. The preform does not flow in the die and only a pressing operation can be carried out. Horizontal surfaces receive press pressure, vertical surfaces get wedge pressure, and wedge pressure means that the felt may be forced to elongate. The completely dry felt is not always much stronger than wet felt. The presence of powdered resin in the preform operates to weaken cellulose fiber-to-fiber bonds. Mosher[24] commented on the difficulties of resin fiber molding:

"Preformed shapes which are relatively flat or have sloping sides can be molded with ease in ordinary all-metal molding dies, and the preformed structure will be retained in the cured piece. If the molds are deep, however, and have steeply sloping sides, the bulk of the preforms generally prevents the use of conventional molding dies, as the preform shears when the male die closes, and the preformed structure is torn and destroyed.

Mosher recommended molding bags to eliminate the difficulty. However, molding bags have not yet proved practical in this type of molding, at least with preforms of high resin content. The bag stretches a good deal at the base, and there is a resin adhesion at that point and an erosion that ruins the bag within a very few pieces. The outlook might be much more promising with low resin preforms.

Because the preform does not flow and pressure distribution during molding is haphazard, the part does not come out of the mold with a good finish. Areas of varying density are present. The problems of finish are exactly those encountered in the die-dried part.

Flowed Resin-Fiber

It may be mentioned that fiber discs of high resin content can be felted, dried, and stacked in the mold to give flowed parts of high finish and excellent physical properties. These, however, amount to molding compounds made by the wet process. They are not pulp moldings in the sense of this discussion. The technique is discussed by Sawyer[25].

Special Preform Treatments

Molded Holes in Felts

In an egg carton a hole for locking the lid is made by fastening a solid block of metal to the screened mold and playing a jet of water on the block as the felter comes out of the slurry. Emery[26] used a tapered upstanding plug set in the screened felting die for the same purpose, and picked the felt up with a transfer having a mating hole. Crabtree[27] set a plug in the felting die mounted on a spring so that the transfer would push it out of the way at the time the felt was taken. Leitzel[28] inset a lift valve in the felting mold and raised it with compressed air to cut through the felt.

Inserts

Screw inserts that are easily molded in flowed plastics have been impossible to put into pulp molding and led to the rather awkward use of rivets for fastening. McNicol[29] solved the problem in a most ingenious manner which, however, only applies to horizontal surfaces. A hollow post is set into the mold so that it is supplied with vacuum. The annular insert is slid down the post until it is stopped and positioned by a collar on the post. A small wire screen is wrapped around the insert and over the end of the post. During felting, the vacuum accretes a felt on the small screen that joins in the main felt. The preform must be moved straight up for release from the mold so that the felt-encapsulated insert slides up the hollow tube. The small screen goes as part of the preform.

Tear Lines

Farnham[34] mounted a blade or a series of blades edgewise on the surface of the felting die, dividing it into a plurality of parts. Felt may form over these blades, but a part line is created that permits easy separation of the preform.

Rolled Edge

Sloan and Sporre[31] rolled the edge of a cloth-covered tropical helmet in a die. Previously sprayed adhesive kept the roll in place. Millions of such dress helmets have been made for the armed services.

Molded Edge

There are several disclosures on thickening the felt at the edge to strengthen it. Randall[02] deliberately felted the felt too long, transferred it in a fitting transfer and dried it in a slightly more shallow die. The extra felt was pushed into place to thicken the edge. The Muller process, as discussed, presqueezes the felt so that it does not elongate during the die dry and the edge can be directly molded.

Break Pattern

DeLuca[33] disclosed a frangible pulp molded nose and tail cone for airborne rocket systems. The drying die is shown in Figure 7-5. It was shaped so as to press a grenade break pattern into the cones. The cones are strong enough to take the aerodynamic loading imposed by the fastest military aircraft. However, when the rockets are fired from within the launcher housing, the fairings are broken, the nose fairing by the passage of the rocket and the tail fairing by the exhaust gases from the rocket. The grenade type construction gives good breakup of the molding and eliminates the danger of fragments impacting and causing damage to the aircraft.

PRODUCTION TECHNIQUES AND PROPERTIES
OF PULP MOLDED PRODUCTS

Disposable Dishes, Egg Cartons and Fruit Packs

These items are felted on automatic machines. The transfer squeezes a portion of the water out and sets the felt in an oven for drying. A form is often used to maintain shape during this operation. The dried part may be humidified and restruck in a die to improve definition. R. W. and J. R. Emery[34] discuss the machinery. A rotary felter might have six egg tray molds set around a cylinder. The cylinder rotates intermittently, with the bottom molds immersed in slurry and felting. One complete revolution is made in 15 to 18 seconds. This gives 20 to 24 molded trays per minute for each row of six molding dies. Machines now in use have from 6 to 16 positions around and from 2 to 10 rows of dies. They produce from 40 to 240 trays per minute.

Pulp for Fruit Packs

Pacific Pulp Molding[35] discussed the pulp used to make fruit and meat trays. The basic raw material was refined groundwood. Wood chips were washed and sent through a single disc refiner at a consistency of 20 to 30%. The pulp was next stored for an hour at 140°F. and sent through a second refiner at 10% consistency. Forty tons of chips were processed per day.

The rotary machines make a dish with bleached fiber on the outside and unbleached inside. Chaplin[36] separated the felting tank into three compartments, with a different pulp in each. The partitions had shaped openings to allow the felting mold to pass. Sealing means were later added to the openings[37]. There is much patent art on the subject, for example, that of Wells[38].

Vases

Flower vases are made on multiple molds lowered on a platform into the pulp vat. The stock used contains a large proportion of waxed paper to give water resistance. The rough bark pattern on the outside is pleasing in white pulp. The contours of a vase are difficult to mold, particularly when the base is wider than another part of the container. Such a molding can be made, however, by placing a premolded porous core of sand on the felting die. Vacuum is applied from the felter through the sand so that it also accumulates a felt. The sand core is left in the bottom of the vase, where it acts as a weight to stabilize it against tipping.

Fusion Felting

Kletzien[39], using two molds, felted the body of the vase and the base separately, then moved the two structures together and continued felting until they joined. Fusion felting was also described by Bushnell[40] and Kyle[41]. The Emerys[34,42] see great possibilities in fusion felting for the production of strong hollow panels for the building trade.

Formed Packaging

Compartmented trays or shaped blocks of pulp can be used to pack delicate parts for shipment. Curtis[43] has written on the subject.

Advertising and Decorative Forms

When the felt is vacuum-accreted inside a screened cavity, all the detail of the mold, including the screen mark, is on the outside of the part. The mold must be separated to obtain the felted piece. In the Lass[44] process large split molds are used to make, for example, advertising signs. When the molds are together, there is a port open at the base. These molds are hung on a large "merry-go-round" type of machine. At the first station the mold is dipped into slurry and pulp is sucked through the port in the bottom to cover the screened inside of the mold. As the mold leaves the slurry, a hose carrying heated air is automatically applied to the open port. The part dries in one revolution of the machine. The split dies open, and an operator removes the part.

Decoy Ducks and Bottles

Felting in split dies is more often done by injecting slurry into the dies under pressure and following this with heated air for at least partial drying. The decoy duck of Figure 7-3 is an example. The slurry enters the dies on the duck's back, leaving a hole that is plugged later.

During the World War II Henkel and Company in Germany made bottles for their chemical products by split-die pressure-felting[45]. Fifty-eight automatic molding machines were operated. The bottle, after being felted under pressure, was dried completely in the mold by heated air. The air was heated by burning gas in a refractory-lined portion of the line. The products of combustion and the heated air were, therefore, used together. This made it necessary to prevent the escape of carbon monoxide. The mold box was designed to form a sealed enclosure around the mold, the outlet being through a hollow base section open only to a vacuum duct.

The pulp was 75% sulfite and 25% mechanical pulp worked up in a beater heated with live steam. Paraffin sizing was used. The cycle to produce a dry part was eighty seconds. At the end of the cycle, the mold opened and the container was blown out on a conveyor.

Radio Speaker Diaphragms

The diaphragm is mounted at its periphery in a rigid basket. The voice coil is glued at the apex of the cone where it operates in a fixed magnetic field. A small corrugated disc, also attached to the basket, keeps the voice coil centered. The pulp molded diaphragm is thinned and flattened near the periphery to provide a hinge for the vibrational excursion. It is thickened towards the apex so as to be able to handle power. The smaller the diaphragm, the poorer its base reproduction is; the larger, the poorer is its treble response. There are, therefore, as many as six speakers in one cabinet.

Diaphragms are held to a weight and resonance specification at the felting tanks. They are felted from as low as 0.1% consistency fiber. The thin wet felt is picked up with a transfer and placed on a form for hot-air pull through drying. The part of the cone that receives the voice coil must be to exact dimension. Kyle[46] provided a guided ring, which slipped over the cone as the drying oven was brought down on it, for this purpose. The cones are dipped into dilute nitrocellulose lacquer for moisture proofing, and may be redipped at the apex for stiffening. Circular cones are placed on a rotating spindle and knife trimmed at apex and periphery. With elliptical cones the periphery is die-cut.

Pulp Molded Filters

The pulp molded mat is most dense on the screen side and most open on the rough bark side. There is a continuous gradation of densities between. When a stream of fluid containing suspended particles is passed through the mat, from bark to screen side, separation and trapping of the particles occur according to their size. This ability to distribute particles allows the depth type filter to collect and hold large amounts of dirt before flow is shut off.

An ingenious high area filter was disclosed by Sloan and Eberman[47]. A stepped cylinder felt was prepared, larger at one end than at the other, and transferred to a folding tool. It was then telescoped while wet to give high area in small volume. The filter has been used for many years in respirators for dusty locations.

Curtis[48] pulp-molded a fuel filter, which was cup-shaped with a mounting rim. He also[49] felted individual filter discs that were locked on a core for oil filtration. In another disclosure[50] he described a

Fig. 7-7. Anderson filter.

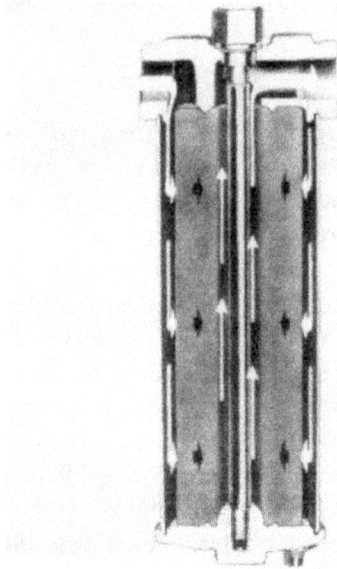

Fig. 7-8. Anderson filter in flow.

thin wall filter containing molded detail to increase filtration area and provide flow channels.

Anderson[51] felted cylindrical depth type filter cartridges as shown in Figures 7-7 and 7-8. These cartridges are resin-impregnated and are self-supporting to over 100 lb/sq in. pressure drop. They may be used for oil or water filtration, and are made in FDA-approved varieties. Curtis[52] formed Anderson's filter cartridge with smooth exterior by raising a shaping tube around the preform during accretion.

Krogel[53] dispersed fibers in acetone or alcohol-water solutions to make depth-type filters of improved flow and dirt-holding capacity. The presence of the solvent prevents the fibers from being over-plasticized by the water; as a result of this, they felt to form far more open and bulky mats. The use of flammable solvents in felting naturally requires extreme care; it also can be quite expensive.

Luggage

Molding of luggage shells has been discussed in an article by Fibre Form, Ltd.[54] Twenty to 30% one-stage phenolic is used in the felt. The material cost is below that for a phenolic molding powder. The suitcase shell weighs one pound and its equivalent weight in phenolic molding powder would be five pounds, without reaching the required strength properties.   The shell is covered with polyvinyl chloride sheeting for finish.

Physical Properties of Molded Luggage

Tensile, lb/sq. in	14 —18,000
Impact 1/2 inch, Izod notched, ft-lb	1.5— 3.0
Shear, lb/sq. in	9 —15,000
Bending, lb/sq. in	15 —19,000
Twenty-four hour water absorption	1 — 10%

Limitations with the technique lie mainly in the fact that moldings must have a certain minimum taper to afford entry of the punch into the loaded die, and also that the molding of inserts cannot at the moment be carried out.   On the other hand, deflashing is not required because moldings are cropped in the tool.

The lower tool of the molding press is heated to 320°F., the upper is run slightly cooler.   Molding pressures are between 500 and 800 lb/sq. in.   An average cycle time for the press operation is two minutes. Specific gravity of the cured material is just over unity.   Inserts for reinforcements cannot be incorporated, nor can the section be appreciably thickened in the areas of higher strength.   The material can be pierced after curing and can be machined if desired.   The surface of the finished molding, though smooth, has a mottled appearance.   There are difficulties in painting because of variations in surface absorbency.

Restaurant Trays

Sawyer[55] described a technique used in remolding in which high finish is obtained on the part from the mold, along with good physical properties.   Total resin content was still low.   The pulp molded preforms were stacked in the die.   The two outside preforms were high in resin, 50 to 60%, to give finish.   The center, thick preform was low in resin, 15 to 30% to give toughness, and for economy.   Molding consolidated the structure. The preforms must be conditioned in a humidity room before molding.   The die clips the preforms so that a molded edge is produced.

Sawyer's method represents an improvement over that of Carter[56], who felted resin-fiber preforms and rubbed them with powdered resin so that they would take a high finish in the remolding operation.

Pulp Moldings for High-Temperature Use

Ceramic fibers such as the aluminum silicates can be formed into useful shapes by pulp molding (Carborundum Company). A convenient method is to slurry the fibers in a stabilized silicon dioxide dispersion. As felted, the ceramic fibers contain enough of the silicon dioxide binder that they dry to hard, high-temperature-resistant materials. The binder dispersion is recycled to the felting tank. Such preforms withstand 2000°F. They furnish a protective layer and shim on steel rod used to tap aluminum furnaces. They function in electric furnances, and as aerospace hardware insulation.

Metal Fiber Specialties

Read, Pollack and McGee[57] discussed forming metal fibers out of slurries. The fibers were dispersed in a viscous fluid and felted. The product was a very porous metal body having a density from 5 to 20% of that of the solid metal. Fiber metal reinforced plastics were made by impregnating porous sintered felts. These have promise for plastic sheet-molding dies of increased life.

Soft metals such as lead or magnesium were reinforced when poured into a strong metal fiber network. An obvious application for a porous metal material is a filter. The pressure drop is low for practical flow rates.

Glass fiber and metal fiber were preformed and then heated to fuse the glass to give a new, low-density cement.

Combustible Cartridge Case

Judge[58] described pulp molding with cellulose nitrate[1] fiber. The cases were felted in a typical pulp molding operation. The moist felt was transferred to a heated die for shaping, presqueezing, and partial drying. The preform was next impregnated with a water-soluble solvent and a curing operation was carried out to bond the resin together and remove the remaining water. The part was then pressed to the final dimensions.

Patents on the process were issued. Beal and Nielsen[59] pulp molded the case from alpha cellulose and die-dried it. They impregnated the case with viscose and regenerated it as a binder. Finally, the case was nitrated in the final form. They also felted the cellulose nitrate fiber as just discussed and mention impregnating with polyvinyl butyral.

DeLuca[60] actually die-dried the cellulose nitrate preform, restricting the temperature of the dies to 250°F., and using extra stabilizer for the cellulose nitrate. He then impregnated with polyvinyl formal solution and baked the excess solvent out in dies under suction, again not exceeding 250°F.

Preforming Glass Fiber by Pulp Molding Method

Glass fiber, drawn fine for strength and treated with a bond promotion agent such as DuPont Volan, is widely used to reinforce polyester or epoxy resin.

The usual preform is made by air deposition. Roving is run through a cutter and sucked onto a shaped screen. A binder is sprayed onto it as it is collected. The screen and preform are dried and baked and the preform is removed for molding. The preform is placed in matched metal trimming dies, a weighed amount of catalyzed liquid polyester resin is poured on, and the press is closed. The final approach of the dies distributes the resin throughout the preform where it cures. If the preform is weak, this sudden flow of resin may tear it. There is little or no strength across such a tear. The preform should be bulky to permit easy impregnation and strong to resist tearing.

Strength in a glass-reinforced plastic molding is nearly directly proportional to the glass content. Glass content is governed by the form the glass is in, as it packs differently in different forms. A glass roving impregnated with polyester resin, to make a fishing rod, may contain as much as 80% glass and show 80,000 lb/sq. in. flexural strength. Woven glass-cloth reinforced plastic may contain 50% glass and show 50,000 lb/sq. in. flexural. The chopped fiber glass preform, air deposited, may have 30% glass content and 30,000 lb/sq. in. flexural strength.

The first attempts to prepare preforms from slurried fiber glass ran into trouble. The cut roving in a stirred slurry separates into strands. Each strand then separates into filaments. While filaments can be accreted, they produce a bulky, cottony felt. When this is dried, molded and the glass content is determined, it is found to be only 3 to 6%. The part is naturally quite weak at this glass content.

Extensive laboratory work showed that the glass, because of the bonding agent with which is had been treated, would wet with and accept polyester resin, which was simply poured into the mixing slurry with it. The sticky polyester resin penetrated to the center of the bundle of filaments in the strand and held them together momentarily. If the resin were catalyzed and the batch heated, the resin was cured and the strand was permanently stabilized. It could be pumped and treated as regular stock. Preforms could be readily made[61] and molded. The stabilized glass strand was found to have high reinforcing value in ordin-

ary pulp moldings and was used to strengthen and toughen pulp-molded luggage.

Weiss and Williams[62] described some of the advantages of the process
as follows: "The preform may be formed in from eight to forty seconds
and is virtually self-supporting as made. It may be removed at once
from the forming screen and transferred to the oven for drying."

The process works well when a small proportion of fiber such as
cellulose fiber is present, as this does not accept the oily resin. Such
fiber acts as a spacer and prevents the sticky glass fibers from agglo-
merating before they are cured[63]. It is important not to "hot-catalyze"
the resin and pour it into a heated slurry. The resin may cure before it
is on the glass and before it impregnates the strand. Unimpregnated
glass with a surface coating of polyester resin separates on pumping
or vigorous stirring. Resin that cures before it is on the glass appears
as small hard spheres. These crush in the molding operation to make
holes and spots. The use of rigid resin allows the fibers to be substantially stiffened so that the desirable bulky preform is produced. Beater
addition methods are used to bind the preform as it dries so that it will
stand up to impregnation at the closing of the press.

The method gives a wide variety of color effects. If the polyester
resin is colored, added to the fiber in the slurry, and left uncured, color
fusion effects are obtained in the molding. If the polyester resin is
colored, added to the fiber, and cured, the fibers are both stabilized and
colored. These colors do not change in the molding operation.[64,65,66]

The wet slurry preforming of glass fiber is used exclusively by the
Cimastra Division of Cincinnati Milling Machine Company[67] and is
described in the catalogue as being the most sophisticated glass-fiber,
reinforced plastic process available. It is used where a showroom appearance as well as complexity of form and great strength are required.
The catalogue states that Cimastra *Wet Slurry* moldings have been used
for high-style modern chairs, bowling settees, dinette chairs, subway seats,
movie projector cases, television cabinets, specialized tool boxes, office
equipment housings, and disc-type children's sleds. According to the
Cimastra sales brochure, automation in felting produces repeated uniformity of product. Coloring of fibers can minimize fiber pattern and
give decorative effects with smoother surface texture. The moldings
have significantly longer wear, better scuff resistance, and less upkeep.

THE FUTURE OF PULP MOLDING

Pulp molding is a true mass production method, operating from a limitless supply of low-priced raw materials. The articles produced are light, strong, and serviceable. They are not temperature-sensitive. They do not become brittle at low temperatures or revert to another shape at high temperatures.

However, finish is a problem, and water absorption is high enough to cause trouble in certain applications. Good finishes can be prepared, but the expense involved makes the article noncompetitive. This situation can be improved by imaginative use of new paints and plastics. Hornbostel's[68] vacuum coating of the preform with plastic sheet in the drying process is an example of what can be done. Water absorption can be controlled at present by high percentage impregnation with polyester resins, but again at too much of an increase in price. Altering the fiber so that it will accept solvent-soluble materials such as varnish in the beater, as discussed, leads to water-stable materials at low cost. A resin should be developed for this process, one that is liquid at the beater preparation temperature of 170°F., but solid and nontacky at felting temperatures.

The success of the presqueezed felt in the Muller technique shows that the die-dry process is still viable. Presqueezing is also valuable for the dry molding process, as water can be pressed out faster than it can be dried, and the presqueeze gives another opportunity to shape the felt.

Pulp molding is ideal for making bucket seats for cars. These require reinforcing with wood or metal to be successful. McNicol's incorporation of inserts into the molding permits such reinforcing to be added in a routine operation.

The record of the three-dimensional fiber industry is one of inventiveness and resourcefulness. The next few years will see new ideas evolved which will invigorate and advance the art, and make the products even more useful than they are at present.

REFERENCES

1. *Papier-Zeitung*, No. 61/62, page 1344, (1939).
2. *Verpackungs Rundshau 7*, No. 5, (1939). (Communication from Harry F. Lewis of the Institute of Paper Chemistry).
3. A. French and C. Frost, U.S. Patent 15228 (1856).
4. E. H. Knight, U.S. Patent 53631 (1866).
5. S. Wheeler and E. Jerome, U.S. Patents 66918 and 66919 (1867).
6. J. Kendall and R. Trested, U.S. Patent 125740 (1872).
7. S. M. Hotchkiss, U.S. Patent 329043 (1885).
8. C. M. Starr, U.S. Patent 429366 (1890).
9. H. Carmichael, U.S. Patent 503738 (1893).

10. M. L. Keyes, U.S. Patent 740023 (1903).
11. G. J. Manson, U.S. Patents 1725465 (1929), 1924409 (1933), 1951940 (1933) and 1983553 (1934).
12. M. P. Chaplin, U.S. Patents 1920292 (1933) and 1988161 (1935).
13. F. Pellegrino, D. Taylor and V. Bernardino, U.S. Patent 3147180 (1964).
14. J. C. Williams, U.S. Patent 3336247 (1967).
15. A. W. Handford, U.S. Patent 1407409 (1922).
16. C. H. Mayne, U.S. Patent 3007842 (1961).
17. F. Louisot, U.S. Patent 1605358 (1926).
18. J. F. Low, U.S. Patent 1680892 (1928)
19. W. Sheffield, U.S. Patent 1984384 (1934).
20. R. J. Box, U.S. Patent 3219520 (1965).
21. C. H. Mayne, U.S. Patent 3306815 (1967).
22. K. Muller and H. Tonniges, U.S. Patent 2841054 (1958).
23. R. U. Haslanger and R. H. Mosher, *Modern Plastics*, *20*, 76 July, (1943). Also see N. J. Taylor, *Plastics*, December (1944); R. H. Mosher and J. B. Griffin, *Modern Plastics*, February, (1945); R. H. Mosher, N. T. Samaras and L. M. Debing, *Paper Trade Journal*, March 15, (1945); and S. H. A. Young, U.S. Patent 2417851 (1947).
24. R. H. Mosher, *Paper Mill News*, April 21, (1945).
25. E. E. Sawyer, U.S. Patent 2343330 (1944).
26. R. L. Emery, U.S. Patent 2754729 (1956).
27. K. L. Crabtree, U.S. Patent 3216890 (1965).
28. A. M. Leitzel, U.S. Patents 2923352 (1960) and 3046187 (1962).
29. J. C. McNicol, U.S. Patent 3106508 (1963).
30. R. A. Farnham, U.S. Patent 2081740 (1937).
31. E. C. Sloan and G. A. Sporre, U.S. Patent 2112384 (1938).
32. W. H. Randall, U.S. Patent 2251243 (1941). Also see Chaplin, U.S. Patent 2377864 (1945).
33. P. L. DeLuca, U.S. Patent 3250839 (1966).
34. R. W. Emery and John R. Emery, *Paper Trade Journal*, 29-33, January 3, 1966.
35. R. P. Hammond, *Paper Trade Journal*, 42-46, September 14, (1964).
36. M. P. Chaplin, U.S. Patent 1701429 (1929).
37. M. P. Chaplin, U.S. Patent 1730450 (1929). Also see U.S. Patents 2995188 (1961) and 3043742 (1962).
38. R. Wells, U.S. Patent 3234080 (1966).
39. H. H. Kletzien, U.S. Patent 2734430 (1956).
40. O. P. Bushnell, U.S. Patent 1690528 (1928).
41. C. Kyle, U.S. Patent 2723600 (1955).
42. R. W. Emery, U.S. Patent 3053728 (1962).
43. R. H. Curtis, U.S. Patent 2416680 (1947).
44. Lass Dehydroform, *Pacific Pulp and Paper*, 13-16, July, 1943. Also see U.S. Patent 2204276 (1940).
45. Paper Pulp Molding Industry in Germany. BIOS Team No. 3258. Technical Information and Document Unit, 38-46 Cadogan Square, London, S.W. 1., distributed by the Office of the Publication Board, Department of Commerce, Washington, D.C.

46. C. Kyle, U.S. Patents 2723423 (1955) and 2804644 (1957).
47. E. C. Sloan and A. H. Eberman, U.S. Patent 2337575 (1943). Also see U.S. Patent 2355714 (1944).
48. R. H. Curtis, U.S. Patent 2663430 (1953).
49. R. H. Curtis, U.S. Patent 2670851 (1954).
50. R. H. Curtis, U.S. Patent 2685969 (1954).
51. L. E. Anderson, U.S. Patents 2539767 (1951) and 2539768 (1951).
52. R. H. Curtis, U.S. Patent 2700326 (1955).
53. G. J. Krogel, U.S. Patent 2802405 (1957).
54. *British Plastics*, 79-81, March (1958). See also *Automobile Engineer*, 407-409, October, (1960).
55. E. E. Sawyer, U.S. Patent 2274095 (1942).
56. W. W. Carter, U.S. Patent 2237048 (1941), reissue 22487 (1944).
57. R. H. Read, W. Pollack and S. W. McGee, *Precision Metal Molding Magazine*, April, (1958).
58. J. F. Judge, *Missiles and Rockets*, January 10, (1966).
59. K. F. Beal and E. R. Nielsen, U.S. Patent 3218907 (1965).
60. P. L. DeLuca, U.S. Patent 3320886 (1967).
61. "Wet Pulping with Stabilized Glass Fiber", *Chemical Processing*, April, (1953).
62. A. C. Weiss and J. C. Williams, *Modern Plastics*, 99-101, November, (1953).
63. D. M. Hawley and J. C. Williams, U.S. Patent 2698558 (1955).
64. D. M. Hawley and J. C. Williams, U.S. Patent 2702241 (1955).
65. D. M. Hawley and J. C. Williams, U.S. Patent 2859109 (1958).
66. D. M. Hawley and J. C. Williams, U.S. Patent 2932601 (1960).
67. Catalogue *FRP*, Cimastra Division, Cincinnati Milling and Machine Company.
68. L. Hornbostel, Jr., U.S. Patent 3205123 (1965).

Index